Study Guide

William Radulovich
Nicholas Belloit

Elementary Algebra

for College Students

Fourth Edition

Allen R. Angel

PRENTICE HALL, Upper Saddle River, NJ 07458

Production Editor: *Carole Suraci*
Acquisitions Editor: *Melissa Acuna*
Supplement Acquisitions Editor: *Audra Walsh*
Production Coordinator: *Ben Smith*

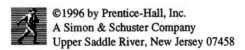
©1996 by Prentice-Hall, Inc.
A Simon & Schuster Company
Upper Saddle River, New Jersey 07458

All rights reserved. No part of this book may be
reproduced, in any form or by any means,
without permission in writing from the publisher.

Printed in the United States of America

10 9 8 7 6 5 4 3

ISBN: 0-13-399221-7

Prentice-Hall International (UK) Limited, *London*
Prentice-Hall of Australia Pty. Limited, *Sydney*
Prentice-Hall Canada Inc., *Toronto*
Prentice-Hall Hispanoamericana, S.A., *Mexico*
Prentice-Hall of India Private Limited, *New Delhi*
Prentice-Hall of Japan, Inc., *Tokyo*
Simon & Schuster Asia Pte. Ltd., *Singapore*
Editora Prentice-Hall do Brasil, Ltda., *Rio de Janeiro*

CONTENTS

Section 1	**1**
Section 2	**37**
Section 3	**75**
Section 4	**101**
Section 5	**145**
Section 6	**177**
Section 7	**215**
Section 8	**277**
Section 9	**303**
Section 10	**341**

Section 1.2 **Fractions**

Summary:

> 1. If a and b represent mathematical quantities, the following may represent the product of a and b: **ab, a b, a(b), (a)b, (a)(b).**
>
> 2. The numbers or variables multiplied in a multiplication problem are called **factors.**
>
> 3. The top number of a fraction is called the **numerator** and the bottom number is called the **denominator.**
>
> 4. A fraction is **reduced to lowest terms** when the numerator and denominator have no factors other than one.

Example 1. In $7(5) = 35$, 7 and 5 are called factors of the product 35.

Example 2. In 9x, 9 and x are factors of the product 9x.

Summary:

> **To Reduce a Fraction to Lowest Terms**
>
> 1. Find the largest number that will divide (without remainder) both the numerator and the denominator. This number is called the **greatest common factor.**
>
> 2. Then divide both the numerator and the denominator by the greatest common factor.

Example 3: Reduce $\frac{24}{40}$ to lowest terms. The common factors of 24 and 40 are 2, 4, and 8. The greatest common factor is 8. Divide both numerator and denominator by 8 to give $\frac{24 \div 8}{40 \div 8}$.

The answer is $\frac{3}{5}$

Example 4. Reduce $\frac{13}{91}$ to lowest terms. Since 13 evenly divides 91, we will divide both numerator and denominator by 13: $\frac{13 \div 13}{91 \div 13} = \frac{1}{7}$

Summary:

> **Multiplication of Fractions:**
>
> Multiply the numerators and place the product over the product of the denominators.

Example 5. Multiply $\frac{6}{11} \cdot \frac{5}{12} = \frac{30}{132} = \frac{5}{22}$. To reduce this fraction, both the numerator and denominator was divided by 6. We could have divided out this common factor of 6 first **before** multiplying.

Example 6. Multiply $\frac{19}{30} \cdot \frac{10}{11}$. By dividing out the common factor of 10 first, we now have $\frac{19}{3} \cdot \frac{1}{11} = \frac{19}{33}$.

Summary:

> **Division of Fractions**
>
> To divide one fraction by another, invert the divisor and proceed as in multiplication.

Division of fractions can be expressed symbolically as follows:

$$\frac{a}{b} \div \frac{c}{d} = \frac{a}{b} \cdot \frac{d}{c} = \frac{ac}{bc}$$

Example 7. $\quad \frac{5}{9} \div \frac{7}{18} = \frac{5}{9} \cdot \frac{18}{7} = \frac{5}{1} \cdot \frac{2}{7} = \frac{10}{7}$ (Notice that we divided out the common factor of 9 before multiplying)

Example 8. $\quad \frac{7}{10} \div 14 = \frac{7}{10} \cdot \frac{1}{14} = \frac{1}{10} \cdot \frac{1}{2} = \frac{1}{20}$ (Once again, notice the common factor of 7 was divided out)

Summary:

> **Addition and Subtraction of Fractions**
>
> To add or subtract two fractions with the same denominator, add or subtract the numerators and put the result over the common denominator.

In symbols, $\quad \frac{a}{c} + \frac{b}{c} = \frac{a+b}{c} \;;\; \frac{a}{c} - \frac{b}{c} = \frac{a-b}{c}$

Example 9. $\quad \frac{1}{12} + \frac{7}{12} = \frac{8}{12} = \frac{2}{3}$. To reduce $\frac{8}{12}$ to lowest terms, the numerator and denominator each was divided by four.

Example 10. $\dfrac{8}{15} - \dfrac{2}{15} = \dfrac{6}{15} = \dfrac{6 \div 3}{15 \div 3} = \dfrac{2}{5}$

Summary:

> To add (or subtract) fractions with different denominators, rewrite each fraction with the same, or common, denominator.

Example 11. $\dfrac{3}{4} + \dfrac{1}{9} = \dfrac{3 \cdot 9}{4 \cdot 9} + \dfrac{1 \cdot 4}{9 \cdot 4} = \dfrac{27}{36} + \dfrac{4}{36} = \dfrac{31}{36}$

Common error: When adding $\dfrac{3}{4} + \dfrac{1}{3}$, students often want to "cancel" the threes. An incorrect answer of $\dfrac{1}{4}$ would be obtained. Dividing out a common factor in the numerator of one fraction and the denominator of another fraction can only be performed when multiplying fractions. See example below:

$\dfrac{3}{4} \cdot \dfrac{1}{3} = \dfrac{1}{4}$ (Divide out common factor of 3)

The following example illustrates the procedure of **changing a mixed number to a fraction.**

Example 12. $7\dfrac{3}{4}$ = what fraction ?

Multiply the denominator, 4, by 7 to obtain 28. To this product, add the numerator of 3. The sum of 31 represents the numerator of the fraction. The denominator of the fraction we are seeking is the same the denominator of the fraction in in the mixed number, 4.

So our final answer is $\dfrac{31}{4}$

The next example illustrates the procedure of **changing a fraction greater than 1 to a mixed number.**

Example 13. Change $\frac{15}{8}$ to a mixed number.

Divide 8 into 15: $8\overline{)15}^{\,1\ r\ 7}$ The quotient, 1, is the whole number part of our mixed number. The remainder, 7, is the numerator of the fraction in the mixed number. The denominator in the fraction of the mixed number will be the same as the denominator of the original fraction. Thus our answer is $1\frac{7}{8}$

The following examples illustrate operations with mixed numbers.

Example 14. $3\frac{3}{4} + 1\frac{1}{2}$

$$= \frac{15}{4} + \frac{3}{2}$$

$$= \frac{15}{4} + \frac{6}{4}$$

$$= \frac{21}{4} = 5\frac{1}{4}$$

Example 15. $2\frac{1}{3} \div \frac{5}{6}$

$$= \frac{7}{3} \div \frac{5}{6}$$

$$= \frac{7}{3} \cdot \frac{6}{5}$$

$$= \frac{7}{1} \cdot \frac{2}{5} \quad \text{(since } 6 \div 3 = 2\text{)}$$

$$= \frac{14}{5} = 2\frac{4}{5}$$

Problem set 1.2

Reduce each fraction to lowest terms:

1. $\dfrac{24}{50}$

2. $\dfrac{91}{13}$

Reduce each fraction to lowest terms:

3. $\dfrac{13}{52}$

4. $\dfrac{70}{35}$

Find the product or quotient:

5. $\dfrac{1}{2} \cdot \dfrac{3}{4}$

6. $\dfrac{12}{5} \div \dfrac{3}{7}$

7. $3\dfrac{1}{4} \div \dfrac{5}{6}$

8. $\dfrac{18}{36} - \dfrac{5}{36}$

9. $\dfrac{1}{3} + \dfrac{2}{5}$

10. $\dfrac{1}{9} - \dfrac{1}{18}$

11. $\frac{5}{6} - \frac{2}{7}$

12. $\frac{4}{5} - \frac{2}{7}$

13. $4\frac{1}{3} - 3\frac{1}{5}$

14. $1\frac{1}{3} + \frac{3}{5}$

15. A board which is 10 $\frac{3}{4}$ feet long is cut into 4 equal lengths. Ignoring the width of each cut, how long is each piece ?

16. A recipe for rice calls for 2 $\frac{1}{4}$ cups of rice. If you wish to make 2 $\frac{1}{2}$ times as much rice, how many cups of rice do you need?

Answers:

1. $\frac{12}{25}$ 2. 7   3. $\frac{1}{4}$ 4. 2   5. $\frac{3}{8}$ 6. $\frac{28}{5} = 5\frac{3}{5}$

7. $\frac{39}{10} = 3\frac{9}{10}$ 8. $\frac{13}{36}$ 9. $\frac{11}{15}$ 10. $\frac{1}{18}$ 11. $\frac{23}{42}$ 12. $\frac{18}{35}$

13. $\frac{17}{15} = 1\frac{2}{15}$ 14. $\frac{29}{15} = 1\frac{14}{15}$ 15. 7 ft. 2 inches or $7\frac{1}{6}$ feet.

16. $5\frac{5}{8}$ cups

1.3 The Real Number System

Summary:

1. A **set** is a collection of **elements** listed within braces.

2. **Natural numbers or counting numbers** include {1, 2, 3, 4,...}.

3. **Whole numbers** include {0, 1, 2, 3, ...}

4. The **positive integers** or **counting numbers** are the same as the **natural numbers**.

5. **Integers** consist of numbers in the set {...-3,-2,-1,0,1,2,3...}

6. **Rational numbers** are numbers which can be expressed as the quotient of two integers, where the denominator of the quotient is not equal to zero.

7. **Irrational numbers** include any numbers that can be represented on a number line which are **not** rational.

8. **Real numbers** are either irrational or rational. All real numbers can be represented on the real number line.

Reminder:
1. All natural numbers are whole numbers.
2. All natural numbers are positive integers.
3. All integers are also rational numbers.
4. All rational numbers are real.
5. <u>No</u> irrational number is rational.
6. All irrational numbers are real.

Example 1. Consider the set

$$\{-\sqrt{3}, -1, -.75, 3, 0, \frac{2}{3}, -3.4, \sqrt{10}\}$$

List elements of the set that are:
a) natural numbers   c) irrational   e) rational
b) integers   d) whole   f) real

Solution:
   a) natural numbers: 3
   b) integers: -1, 3, 0
   c) irrational: $-\sqrt{3}$, $\sqrt{10}$
   d) whole: 0, 3
   e) rational: -1, -.75, 3, 0, 2/3, -3.4
   f) real: $-\sqrt{3}$, -1, -.75, 3, 0, $\frac{2}{3}$, -3.4, $\sqrt{10}$

Exercise set 1.3

List elements of each of the following sets.

1. Natural numbers > 7

2. Integers between -5 and 1

3. Whole numbers between 3 and 4

4. Integers between -4 and 5

5. Positive integers less than 7

True or Fale:

6. $\frac{1}{3}$ is an integer.   7. -2 is a rational number.

8. .75 is a rational number.   9. $\sqrt{9}$ is an irrational number.

10. $\sqrt{7}$ is real.   11. all integers are real numbers.

12. No whole number is rational.

13. Some real numbers are not rational.

14. Consider $s = \{ -\frac{2}{3}, 0, -3, 4, 7\frac{1}{2}, \sqrt{5}, -\sqrt{11}, 1.29, 308 \}$

   List those numbers that are:

   a) positive integers
   b) whole numbers
   c) integers
   d) rational numbers
   e) real numbers

Answers to Exercise set 1.3

1. {8, 9, 10, 11, ...}   2. {-4, -3, -2, -1, 0}

3. ∅ null set   4. {-3, -2, -1, 0, 1, 2, 3, 4}

5. {..., 0, 1, 2, 3, 4, 5, 6}   6. False

7. True   8. True

9. False   10. True

11. True   12. False

13. True

14. a) {4, 308}
    b) {0, 4, 308}
    c) {0, -3, 4, 308}
    d) $\{ -\frac{2}{3}, 0, -3, 4, 7\frac{1}{2}, 1.29, 308 \}$

    e) $\{ -\frac{2}{3}, 0, -3, 4, 7\frac{1}{2}, \sqrt{5}, -\sqrt{11}, 1.29, 308 \}$

Section 1.4   Inequalities

Summary:

> 1. The number line is used to explain inequalities.
>
> 2. The number to the right on a number line is the greater number. The number to the left is the smaller number.
>
> 3. The symbol for "greater than" is >.
>
> 4. The symbol for "less than" is <.
>
> 5. The point of the inequality symbol always "points" to the smaller number.

Example 1.   Insert a > or < symbol between numbers to make a true statement.

			Answers
a)	-5	7	-5 < 7
b)	$\frac{1}{3}$	$\frac{1}{4}$	$\frac{1}{3} > \frac{1}{4}$
c)	-55	-62	-55 > -62
d)	$-2\frac{1}{2}$	-2.75	$-2\frac{1}{2} > -2.75$

Summary:

> 1. The absolute value of a number is the distance between the number and zero on a real number line.
>
> 2. The absolute value of every number is either positive of zero.
>
> 3. The negative of the absolute value of a nonzero number will always be a negative number.

Example 2. Insert >, <, or = between the two numbers to make a true statement.

			answers
a)	$|4|$	4	=
b)	-2	$|-3|$	<
c)	-3	$-|-3|$	=
d)	$|-6|$	0	>
e)	$-|-18|$	$-|-17|$	<

Exercise Set 1.4

Insert a > or < between the numbers to make a true statement:

1. -3.4 -3

2. $-\frac{2}{3}$ -1

3. $\frac{7}{4}$ $\frac{4}{7}$

4. $|-1.9|$ $|-2.3|$

5. $-|2.3|$ $-(-4.9)$

6. $\left|-\frac{7}{2}\right|$ $\left|-\frac{2}{7}\right|$

7. $\frac{3}{4} \cdot \frac{3}{4}$ $\frac{3}{4} + \frac{3}{4}$

8. $3\frac{1}{3} \cdot 2$ $3\frac{1}{3} + 2$

9. $\dfrac{7}{8} - \dfrac{1}{2}$ $\dfrac{7}{8} \div \dfrac{1}{2}$ 10. $-|-34|$ $-|-33|$

11. -0.007 $-.008$ 12. 12 $\left|\dfrac{-25}{2}\right|$

Answers to exercise set 1.4

1. <	2. >	3. >	4. <
5. <	6. >	7. <	8. >
9. <	10. <	11. >	12. <

Section 1.5 Addition of Real Numbers

Summary:

> 1. You can use a number line to add signed numbers.
>
> 2. Any 2 numbers whose sum is zero are said to be opposites of each other.
>
> 3. **To add real numbers with the same sign**, add their absolute values. The sum has the same sign as the numbers being added.
>
> 4. **To add two real numbers with different signs**, find the difference between the larger absolute value and the smaller absolute value. The answer has the sign of the number with the larger absolute value.

Example 1.   Add -5 + 8 using a number line.

$$\overset{-5}{\longleftarrow}\quad\overset{+8}{\longrightarrow}$$
$$\longleftarrow\!-\!|\!-\!-\!-\!-\!-\!|\!-\!-\!-\!|\!-\!-\!-\!\longrightarrow$$
$$\quad\quad -5\quad\quad 0\quad 3$$

Example 2.   Add (-7) + (-4) using a number line.

$$\overset{-4}{\longleftarrow}\overset{-7}{\longleftarrow}$$
$$\longleftarrow\!|\!-\!-\!-\!-\!-\!-\!-\!|\!-\!-\!-\!-\!-\!\longrightarrow$$
$$\quad -11\quad\quad\quad 0$$

Example 3.   Add -6 + 6 = 0 using a number line.

$$\overset{+6}{\longrightarrow}$$
$$\longleftarrow\!-\!|\!-\!-\!-\!-\!-\!|\!-\!-\!-\!-\!-\!-\!\longrightarrow$$
$$\quad\quad -6\quad\quad 0$$

Since -6 + 6 = 0, we say that -6 and 6 are <u>opposites</u> of each other.

Example 4.   Find the opposite of each of the following numbers:

   a)   -3, answer is 3 since -3 + 3 = 0

   b)   4, answer is -4 since -4 + 4 = 0

Example 5.   Add 5 + 9

Since both 5 and 9 have same sign we add their absolute values.
$|5| + |9| = 5 + 9 = 14$

Example 6. Add -7 + -14

Since both numbers have the same sign we add
$|-7| + |-14| = 7 + 14 = 21$
Since both numbers being added are negative, their sum is negative so -7 + -14 = -21.

Example 7. Add 12 + (-7)

1. Find the difference between larger absolute value and smaller
$|12| - |-7| = 5$

2. The answer is 5 since the sign of the answer is the same as the sign of the number with the larger absolute value.

Example 8. Add -13 + 6

1. Find the difference between $|-13|$ and $|6|$ = 13 - 6 = 7
2. The answer is -7. The sign of the answer is negative since the sign of the number with the larger absolute value is negative.

Remember: The sum of 2 signed numbers with different signs may be either positive or negative. The sign of the sum will be the same as the sign of the number with the larger absolute value.

Exercise Set 1.5

Add as indicated:

1. 10 + 5   2. 10 + (-7)   3. -7 + 12

4. -12 + 5   5. 50 + (-35)   6. (-34) + (-13)

7. -103 + 73   8. -253 + (-147)   9. 59 + -83

10. 0 + (-43)   11. 12 + (-97)   12. -23 + (-59)

13. -138 + 600   14. -0 + 85   15. -0 + -0

16. -15 + 15   17. -19 + 19   18. -56 + 56

19. A football player gained 18 yards on one play but lost 12 yards on the second play. What is the total net yardage gained or lost?

20. A total of $622 of checks were written on a checking account. Then $1000 was deposited. What is the net change in the bank balance?

Answers to exercise set 1.5

1.	15	2.	3	3.	5	4.	-7
5.	15	6.	-47	7.	-30	8.	-400
9.	-24	10.	-43	11.	-85	12.	-82
13.	462	14.	85	15.	0	16.	0
17.	0	18.	0	19.	6	20.	$378

Section 1.6   **Subtraction of real numbers**

Summary:

> If a and b represent any two real numbers, then
>
> $$a - b = a + (-b)$$
>
> In other words, to subtract b from a , add the **opposite** of b to a

Example 1.   8 - 3 =
             8 + (-3) = 5

Example 2.   5 - 11 =
             5 + (-11) =
             (Think: |-11| - |5| = 11 - 5 = 6, answer is -6)
             So 5 + -11 = -6

Example 3.   -9 - (3) =
             -9 + (-3) = -12

Example 4.   -9 - (-7) =
             -9 + (+7) = -2

Example 5.   Mark's checkbook balance was $235. He then wrote a check for $380, what is the new checkbook balance?

             235 - 308
             235 + (-380) = $-145

In evaluating expressions involving more than one addition or subtraction, proceed from left to right unless grouping symbols are present.

Example 6.   -6 - 8 + 15 =
             -6 + (-8) + 15 =
               -14   + 15 = 1

Example 7.   -9 - (-7) + 8 - (17) =
             -9 + (+7) + 8 + (-17) =
                -2   + (-17)   = -11

Exercise Set 1.6

Evaluate the following:

1. 7 - 4
2. -6 - 10
3. 5 - 5
4. -5 - (-9)
5. 8 - 17
6. 9 - (-14)
7. -20 - (-11)
8. 20 - (-9)
9. -20 - 9
10. (-7) - (-12)
11. -25 - 17
12. 25 - 48
13. 17 - 17
14. Subtract 8 from 19
15. Subtract -2 from -15
16. -8 + 10 - 9
17. 17 + (-13) - 10 + 18
18. 19 - 20 - 18 - 17
19. The temperature in Detroit, Michigan fell from 32°F to -23°F. How much did the temperature drop?
20. An airplane is 2300 ft above sea level while a submarine is 1250 ft below sea level. How far above the submarine is the airplane?

Answers to exercise set 1.6

1. 3
2. -16
3. 0
4. 4
5. -9
6. 23
7. -9
8. 29
9. -29
10. 5
11. -42
12. -23
13. 0
14. 11
15. -13
16. -7
17. 12
18. -36
19. 55°
20. 3550 ft

Section 1.7   **Multiplication and division of real numbers**

Summary:

> 1. The product of 2 real numbers with **like signs** is a **positive number**.
>
> 2. The product of 2 real numbers with **unlike signs** is a **negative number**.
>
> 3. Zero times any number equals 0.
>
> 4. The product of an **even** number of negatives will always be **positive**.
>
> 5. The product of an **odd** number of negatives will always be **negative**.

Example 1. $(-5)(-4) = +20$ or 20 (like signs, product positive)

Example 2. $(-8)(14) = -112$ (unlike signs, product is negative)

Example 3. $(12)(15) = 180$ (like signs, product is positive)

Example 4. $0 \cdot 56 = 0$ (zero multiplied by **any** real number is always zero.)

Example 5. $\left(\dfrac{-5}{9}\right)\left(\dfrac{3}{11}\right) = \dfrac{(-5)(3)}{(9)(11)} = \dfrac{-5}{3 \cdot 11} = -\dfrac{5}{33}$

Example 6. $(-5)(8)(-3)(4) =$
             $(-40)(-12)   = +480$

   Note: The product of an even number of negatives will always be positive.

Example 7. $(-3)(-2)(-5)(-8)(-1) =$
   Think $(3)(2)(5)(8)(1)   = 240$
   Answer is -240 since the product of an **odd** number of negatives will always be negative.

20

Summary:

1. The quotient of 2 real numbers with **like signs** is a **positive number**.

2. The quotient of 2 real numbers with **unlike signs** is a **negative number**.

Example 8. $-35 \div 5 = \dfrac{-35}{5} = -7$

Unlike signs; answer is negative.

Example 9. $\dfrac{-45}{-9} = 5$

Since -45 and -9 have the same sign, the quotient is positive.

Example 10 $\dfrac{-3}{4} \div \dfrac{-5}{8} = \dfrac{-3}{4} \cdot \dfrac{8}{-5} = \dfrac{-6}{-5} = \dfrac{6}{5}$

Note that $\dfrac{8}{-5} = \dfrac{-8}{5} = -\dfrac{8}{5}$

Summary:

1. For all a, b, b ≠ 0,
$\dfrac{a}{-b} = \dfrac{-a}{b} = \dfrac{a}{b}$.

Example 11. $\dfrac{4}{9} \div \dfrac{-5}{27} =$

$\dfrac{4}{9} \cdot \dfrac{27}{-5} = \dfrac{4(27)}{9(-5)} = \dfrac{12}{-5} = \dfrac{-12}{5}$

It is customary to write the negative symbol either in the numerator or in front of the fraction.

Summary:

1. $\dfrac{0}{a} = 0$, $a \neq 0$

2. $\dfrac{a}{0}$ is undefined

Example 12. $\dfrac{0}{-17} = 0$ since $\dfrac{0}{a} = 0$ for all $a \neq 0$.

Example 13. $\dfrac{-17}{0}$ is undefined since $\dfrac{a}{0}$ is undefined for all $a \neq 0$.

Exercise Set 1.7

Find the product:

1. $8(-3)$
2. $(-17)(-9)$
3. $-1(8)$
4. $2(-4)(-9)$
5. $(-1)(-1)(-1)(17)$
6. $(-1)(50)(0)$
7. $(5)(-3)(-4)(-2)$
8. $\left(\dfrac{-5}{7}\right)\left(\dfrac{-7}{9}\right)$
9. $\left(\dfrac{-3}{4}\right)\left(\dfrac{2}{3}\right)$
10. $\left(\dfrac{-5}{7}\right)\left(\dfrac{-6}{8}\right)$
11. $\left(\dfrac{-9}{10}\right)\left(\dfrac{0}{5}\right)$

Find the quotient:

12. $\dfrac{-18}{3}$
13. $-25 \div (-5)$
14. $12 \div (-4)$
15. $\dfrac{-5}{5}$
16. $\dfrac{-91}{-13}$
17. $\dfrac{-39}{52}$

18. $\dfrac{0}{50}$ 19. $\dfrac{-20}{0}$ 20. $-\dfrac{16}{3} \div \dfrac{-5}{9}$

21. $\dfrac{7}{15} \div \left(\dfrac{-9}{10}\right)$ 22. $0 \div 7$ 23. $7 \div 0$

Answers to exercise set 1.7

1. -24 2. 153 3. -8 4. 72

5. -17 6. 0 7. -120 8. $\dfrac{5}{9}$

9. $-\dfrac{1}{2}$ 10. $\dfrac{15}{28}$ 11. 0 12. -6

13. 5 14. -3 15. -1 16. 7

17. $\dfrac{-3}{4}$ 18. 0 19. undefined

20. $\dfrac{48}{5} = 9\dfrac{3}{5}$ 21. $\dfrac{-14}{27}$ 22. 0

23. undefined

Section 1.8   **An Introduction to Exponents**

Summary:

---
1. $a \cdot a = a^2$, $a \cdot a \cdot a = a^3$ and in general $a \cdot a \cdot a \cdot \ldots = a^n$ n factors of a

2.   a is called the **base** and n is called the **exponent**.

Example 1.   Evaluate each expression

a) $4^2 = 4 \cdot 4 = 16$

b) $(-3)^3 = (-3)(-3)(-3)$
    $= (9)(-3) = -27$

c) $\left(\frac{3}{4}\right)^2 = \frac{3}{4} \cdot \frac{3}{4} = \frac{3 \cdot 3}{4 \cdot 4} = \frac{9}{16}$

d) $1^4 = 1 \cdot 1 \cdot 1 \cdot 1 = 1$

You can write algebraic expressions using exponential notation:

Example 2.

a) $x \cdot x \cdot x = x^3$

b) $a \cdot a \cdot b \cdot b \cdot b = a^2 \cdot b^3$

c) $3 \cdot 3 \cdot 3 \cdot (ab)(ab) = 3^3 \cdot (ab)^2$

d) $w \cdot w \cdot y = w^2 y^1 = w^2 y$

Whenever we see a variable raised to an exponent of 1, we can write the variable without the exponent for simplicity.

Example 3.

$x^1 \cdot y^1 = xy$

Summary:

> An exponent refers to only the number or letter that directly precedes it unless parentheses are used to indicate otherwise.

Example 4. Evaluate the following.

a) $-2^4 = -1 \cdot (2^4) = -1 \cdot 16 = -16$

b) $(-2)^4 = (-2)(-2)(-2)(-2) = 16$

c) $-a^3 = -1 \cdot (a^3)$ (only the a is cubed not negative a)

Example 5. Evaluate x^4 and $-x^4$ for $x = -2$

If $x = 2$ then $x^4 = 2^4 = 2 \cdot 2 \cdot 2 \cdot 2 = 16$

If $x = -2$ then $-x^4 = -1 \cdot (x^4) = -1 \cdot (-2)^4 = -1 \cdot 16 = -16$

Exercise Set 1.8

Evaluate each expression

1. 3^4
2. 1^5
3. $(-2)^3$
4. -7^2
5. -9^2
6. $(-9)^2$
7. $(-1)^3(4^2)$
8. $(-2)^3(-1)^2$
9. $(-2)^4(-1)^3$
10. $-2^4(-1)^2$
11. $4(-5)^2$

Express in exponential form:

12. $a \cdot a \cdot a \cdot a \cdot b \cdot b \cdot b$
13. $\left(\frac{1}{2}\right)\left(\frac{1}{2}\right)\left(\frac{1}{2}\right)(x \cdot x)$
14. $x \cdot y \cdot y \cdot y \cdot x \cdot y \cdot y \cdot x$
15. $3\ aa \cdot 3 \cdot a \cdot 3b \cdot 3b$

Express as a product of factors:

16. x^3y^2
17. $3^3 \cdot a^2b^4$
18. $(-2)^2y^4 \cdot z$
19. $(-1)^3a^3b^2$

Evaluate (a) $-x^2$ and (b) x^2 for each of the following values of x:

20. -3   21. 2   22. -7   23. 0

24. $-\dfrac{1}{3}$ 25. $\dfrac{3}{5}$

Answers to exercise set 1.8

1. 81   2. 1   3. -8   4. -49
5. -81   6. 81   7. -16   8. -8
9. -16   10. -16   11. 100   12. a^4b^3
13. $\left(\dfrac{1}{2}\right)^3 \cdot x^2$ 14. $x^3 \cdot y^5$ 15. $3^4 a^3 b^2$

16. x · x · x · y · y
17. 3 · 3 · 3 · a · a · b · b · b
18. (-2)(-2)(y)(y)(y)(y)(z)
19. (-1)(-1)(-1) · a · a · a · b · b
20. a)   -9   b)   9
21. a)   -4   b)   4
22. a)   -49   b)   49
23. a)   0   b)   0
24. a) $-\dfrac{1}{9}$ b) $\dfrac{1}{9}$

25. a) $-\dfrac{9}{25}$ b) $\dfrac{9}{25}$

Section 1.9   **Use of Parentheses and Order of Operations**

Summary:

> To evaluate mathematical expression, the following order is used:
>
> 1. Evaluate expressions within **parentheses** first.
>
> 2. Next, evaluate all **exponential** expressions.
>
> 3. Next, evaluate all **multiplications** and **divisions** in order, working from left to right.
>
> 4. Finally, evaluate all **additions** and **subtractions** in the order in which they occur, working from left to right.

Example 1. $3 + 4^2 \cdot 5 - 7 =$
            $3 + 16 \cdot 5 - 7 =$ (exponents first)
            $3 + 30 - 7 \; =$ (multiplication next)
            $83 - 7 =$
            76

Example 2. $(3 + 4^2) \cdot 5 - 7 =$
            $(3 + 16) \cdot 5 - 7 =$ (exponents within parentheses first)
            $19 \cdot 5 - 7 =$ (do work within parentheses
            $95 - 7 =$ (multiplication before subtraction)
            88

Example 3. $(9 \div 3) + 5(2 - 4)^2 =$
            $(3) + 5(-2)^2$ = (parentheses first)
            $3 + 5(+4)$ = exponents, next
            $3 + 20$ = multiplication before addition
            23

Example 4. $-5^2 + 8 \div 4 =$
            $-(5)^2 + 8 \div 4 =$ (rewrite, remember that 5 is the base,
                                      not negative 5.)
            $-25 + 8 \div 4$ = (exponents first)
            $-25 + 2$ = (division next)
            -23

Example 5. $(-5)^2 + 8 \div 4 =$ (here, -5 is the base)
$25 + 8 \div 4\quad =$ (exponents first)
$25 + 2\qquad\quad =$ (division next)
27

Example 6. $-9 - 18 \div 9 \cdot 3^2 + 6\ =$
$-9 - 18 \div 9 \cdot 9 + 6\ =$ (exponent)
$-9 - (18 \div 9) \cdot 9 + 6 =$ (division and multiplication
$\qquad\qquad\qquad\qquad\qquad\qquad$ next from left to right)
$-9 - (2) \cdot 9 + 6\qquad =$
$-9 - 18 + 6\qquad\qquad =$
$-9 + (-18) + 6\qquad\ \ =$
$-27 + 6\qquad\qquad\quad\ =$
-21

Example 7. $\dfrac{5}{9} - \dfrac{3}{3} \cdot \dfrac{4}{7}$

$\dfrac{5}{9} - \left(\dfrac{3}{4} \cdot \dfrac{4}{7}\right)\qquad$ Multiplication first

$\dfrac{5}{9} - \dfrac{3}{7}\qquad$ LCD is $9 \cdot 7 = 63$

$\dfrac{5}{9} \cdot \dfrac{7}{7} - \dfrac{3 \cdot 9}{7 \cdot 9}\qquad$ Rewrite fractions using LCD

$\dfrac{35}{63} - \dfrac{27}{63}$

$\dfrac{8}{63}$

Example 8. Evaluate $8y - 17$ for $y = -2$

1. First, substitute -2 for y:
   $8(-2) - 17$
2. Next, perform multiplication first
   $-16 - 17$
3. $-16 + -17 = -33$

Example 9. Evaluate $-x^2 + 2(y - 1) + 7$, when $x = -3$ and $y = 2$

    1. Substitute values for variables

$$-(-3)^2 + 2(2 - 1) + 7 =$$
$$-(-3)^2 + 2(1) + 7 =$$
$$-(9) + 2(1) + 7 =$$
$$-9 + 2 + 7 =$$
$$0$$

Exercise Set 1.9

Perform the operations and simplify

1. $5 + 2(3)$ 2. $(7^2 \cdot 3) - (5 - 4)$

3. $[12 - (6 \div 3)] - 6$ 4. $3^2 - 5^2(5 - 2)^2$

5. $-3^2 + 4 \cdot 9$ 6. $(4^2 - 1) \div (4 + 1)^2$

7. $2.5 + 7.5 \div .3 + (.5)^2$ 8. $2(5.3) + (4.3)^2 - 3.05$

9. $\dfrac{5}{8} - 5 \cdot \dfrac{1}{8}$ 10. $3\left(\dfrac{4}{5} + 4\right) \div \left(\dfrac{2}{5}\right)^2$

11. $(8 + 5)^2 - (2 - 4)^2$

Write the following statement as a mathematical expression using parentheses and brackets and then evaluate:

12. Multiply 7 by 3 and to this product, add 15. Divide the sum by 8. Multiply the quotient by 9.

Evaluate the given expression for the indicated values:

13. $3x + 6$ for $x = -2$ 14. $-2x^2 + 3x + 4$ for $x = -3$

15. $4(x - 3)^2$ for $x = 5$ 16. $5x + 2y^2 - 6$ for $x = 2$, $y = -3$

17. $3x^2 - 2y^2 - 6$ for $x = 2$, $y = -4$

18. $-2x^2 - 4y^2 + 6(y + 1)$ for $x = -1$, $y = 2$

19. $4x^2 + 5x - 3$ for $x = -2$

20. $\dfrac{7x^2}{3} + \dfrac{3x^2}{2}$ for $x = 2$ 21. $5x^2 + 2xy - y^2$

                                                                                    for $x = -3$, $y = 2$

Answers for exercise set 1.9

1. 11	2. 146	3. 3	4. −216

5. 27	6. $\frac{3}{5}$	7. 27.75	8. 26.04

9. 0	10. 90	11. 165	12. $\frac{81}{2}$

13. 0	14. −23	15. 16	16. 22

17. −20	18. 0	19. 3	20. $\frac{46}{3}$

21. 29

Section 1.10   **Properties of the Real Number System**

Summary:

> 1. **The commutative property of addition** states: If a and b are 2 real numbers then
>
> a + b = b + a.
>
> 2. **The commutative property of multiplication** states: If a and b are 2 real numbers, then a · b = b · a.
>
> 3. The commutative property **does not** hold for subtraction or division.
>
> 4. The **associative property of addition** states: If a, b and c represent any 3 real numbers, then
>
> (a + b) + c = a + (b + c).
>
> 5. The **associative property of multiplication** states: If a,b and c are any 3 real numbers, then
>
> (a · b) · c = a · (b · c).
>
> 6. The **distributive property** states: If a, b and c represent any 3 real numbers, then
>
> a(b + c) = ab + ac.

Hints:

1. The commutative properties involve changes in order while the associative properties involve changes in grouping.
2. The distributive property involves 2 operations: Multiplication and addition.

For each of the examples below, name the property involved.

		Answers

Example 1. $3(x + 4) = 3x + (3 \cdot 4)$ Distributive property

Example 2. $5 + (-3) = -3 + 5$ Commutative property of addition

Example 3. $(5 \cdot 7) \cdot x = 5 \cdot (7 \cdot x)$ Associative property of multiplication

Example 4. $(x + 3) + 7 = (3 + x) + 7$ Commutative property of addition

Example 5. $5x + 6x = (5 + 6)x$ Distributive property (in reverse order)

Example 6.   Name the property used to go from one step to the next.

$8 + 5(x + 3) =$
$8 + 5x + 5(3) =$ **Distributive**
$8 + 5x + 15 =$ **Arithmetic fact**
$8 + 15 + 5x =$ Commutative property of addition
$(8 + 15) + 5x$
$23 + 5x$ Addition fact
$5x + 23$ Commutative property of addition

Example 7.   Name the property to go from one step to the next:

$-1(3x + 2) + 2(3x + 6) =$
$-1(3x) + -2(2) + 2(3x) + 2(6)$ Distributive property
$(-1 \cdot 3)x + -1(2) + (2 \cdot 3)x + 2 \cdot 6$ Associative property of Multiplication
$-3x + (-2) + 6x + 12$ Arithmetic facts
$-3x + 6x + (-2) + 12$ Commutative property of addition
$(-3 + 6)x + (-2) + 12$ Distributive property
$3x + 10$ Arithmetic facts

Exercise Set 1.10

Name the property illustrated:

1. $4(2 + 7) = 4(2) + 4(7)$
2. $x \cdot a = a \cdot x$
3. $-1(x + 5) = -1 \cdot x + (-1)(5)$
4. $x + (2 + 3) = (x + 2) = 3$
5. $(x \cdot 2) \cdot 3 = x \cdot (2 \cdot 3)$
6. $(a + b) + x = (b + a) + x$

Name the property illustrated.

7. $2(3x + 4y) = 2(3x) + 2(4y)$

Complete each exercise using the given property.

8. $x + 3$ 9. $2(3x + 5)$ 10. $(x + 3) + 9$

Name the property used to go from one step to the next.

11. $8 + 3(2x + 4) =$
   a) $8 + 3(2x) + 3(4)$
   b) $8 + (3 \cdot 2)x + 3(4)$
   c) $8 + 6x + 12$
   d) $8 + 12 + 6x$
   e) $20 + 6x$
   f) $6x + 20$

Answers to exercise set 1.10

1. Distributive property
2. Commutative property of multiplication
3. Distributive property
4. Associative property of addition
5. Associative property of multiplication
6. Commutative property of addition
7. Distributive property

8. 3 + x   9. 2(3x) + 2(5)   10. x + (3 + 9)

11. a. Distributive property
    b. Associative property of multiplication
    c. Arithmetic fact
    d. Commutative property of addition
    e. Arithmetic fact
    f. Commutative property of addition

Chapter 1 Practice Test

1. Consider the set of numbers

    $S = \{-7, 58, -2\frac{1}{2}, 0, 4.58, \sqrt{7}, \frac{7}{8}, -3, -10, \sqrt{9}\}$

    List those that are:
    a. natural numbers   b. whole numbers
    c. integers   d. rational numbers
    e. irrational numbers   f. real numbers

2. True or false: Every integer is a rational number.

3. True or false: All whole numbers are also natural numbers.

Insert a >, <, = in the blank to make each statement true.

4. -4 ____ -5   5. -|-4| ____ -|-5|

6. -2^2 ____ 4   7. |3 - 7| ____ |7 - 3|

Evaluate each expression:

8. 2 · 3 + 4   9. -2 - 4 - 6

10. (-4)(3)(-2)(-2)   11. $\left(-14 \div \frac{1}{7}\right) - (-2)$

12. $2 \cdot 3^2 - 4 \cdot 5^2$ 13. $\left(\frac{-3}{4}\right)^2$

14. -7(-2 - 5) ÷ 7   15. -9.1 ÷ .13

16. $(-2)^5$

17. Write 3 · 3 · 5 · 5 · a · a · bbb in exponential form.

18. Write $3^3 \cdot 4^2 \cdot u^4 v$ as a product of factors.

Evaluate each expression for the indicated values.

19. $3x^2 - 4x$ for $x = -2$ 20. $-x^2 + 3x + 7$ for $x = 3$

21. $3x - 2y^2 + 4$ for $x = -3, y = -2$

Name the property illustrated (22 - 25).

22. $2(3x + 2) = 2(3x) + 2(2)$ 23. $2 + 3y = 3y + 2$

24. $2(3 \cdot z) = (2 \cdot 3)z$ 25. $3 + (4 + y) = (3 + 4) + y$

Answers to Chapter one Practice test

1. a) $58, \sqrt{9}$ b). $58, 0, \sqrt{9}$ c). $-7, 58, 0, -3, -10, \sqrt{9}$

   d). $-7, 58, -2\ 1/2, 0, 4.58, 7/8, -3, -10, \sqrt{9}$

   e). $\sqrt{7}$ f) $-7, 58, -2\ 1/2, 0, 4.58, \sqrt{7}\ 7/8, -3, -10, \sqrt{9}$

2. True   3. False

4. >   5. >   6. <   7. =

8. 10   9. -12   10. -48   11. -96

12. -82   13. 9/16   14. 7   15. -70

16. -32   17. $3^2 5^2 a^2 b^3$ 18. $3 \cdot 3 \cdot 3 \cdot 4 \cdot 4 \cdot u \cdot u \cdot u \cdot u \cdot v$

19. 20   20. 7   21. -13

22. distributive property   23. commutative property of addition

24. associative property of multiplication   25. associative property of addition

Section 2.1 - Combining Like Terms

Summary:

> **Variables** are letters used to represent numbers.
>
> An **expression** (or **algebraic expression**) is a collection of numbers, variables, grouping symbols, and operation symbols.
>
> **Terms** refer to the parts of an algebraic expression that are added or subtracted.
>
> The **coefficient** (or **numerical coefficient**) is the numerical part of a term. If a term appears without a numerical coefficient, we assume that the coefficient is 1.
>
> A **constant** (or **constant term**) is one which contains no variables, consisting only of a number.
>
> **Like terms** are terms that have the same variables with the same exponents.

Example 1. Identify any like terms in the following algebraic expressions.

a) $5y - x + 3y$ $5y, 3y$ are like terms

b) $4x^3 + x + x^2 + x^3$ $4x^3, x^3$ are like terms

c) $ab^2 + ab + 5ab^2 - 2ab^2$ $ab^2, 5ab^2, -2ab^2$ are like terms

d) $x^2 + 2x + y^2$ no like terms

Summary:

> **To combine like terms**
>
> 1. Determine which terms are like terms.
>
> 2. Add or subtract the **coefficients** of like terms
>
> 3. Multiply the number found in step 2 by the common variables.

Example 2. Combine like terms: $5x + 4x$

    a) Determine like terms $5x, 4x$
    b) Add coefficients of like terms $5 + 4 = 9$
    c) Multiply the number found in (b) by the common variables. Therefore, $5x + 4x = 9x$

Example 3. Combine like terms: $\frac{3}{4}x - \frac{5}{6}x$

    a) Determine like terms $\frac{3}{4}x, \frac{5}{6}x$

    b) Subtract coefficients of like terms

$$\frac{3}{4} - \frac{5}{6} = \frac{9}{12} - \frac{10}{12} = \frac{-1}{12}$$

    c) Multiply the number found in (b) by the common variables. Therefore, $\frac{3}{4}x - \frac{5}{6}x = -\frac{1}{12}x$

Example 4. Combine like terms: $4y + 7 - 7y$

    We can rearrange as $4y - 7y + 7$
    a) Determine like terms $4y, 7y$
    b) Subtract coefficients of like terms $4 - 7 = -3$
    c) Multiply the numbers found in (b) by the common variables. Therefore, $4y + 7 - 7y = -3y + 7$

Example 5. Combine like terms: $2x + 4y + 6 - 3y + 7x - 8$

Again, rearrange as $2x + 7x + 4y - 3y + 6 - 8$
a) Determine like terms $2x, 7x$
$4y, 3y$
$6, 8$
b) Add or subtract the coefficients of like terms
For x: $2 + 7 = 9$
For y: $4 - 3 = 1$
For constants: $6 - 8 = -2$
c) Multiply the numbers found in step 2 by the common variables. Therefore,
$2x + 4y + 6 - 3y + 7x - 8 = 9x + y - 2$

Summary:

Distributive Property

For any real numbers **a, b, c,**

a(b + c) = ab + ac.

Example 6: Use the distributive property to remove parentheses

a) $3(x + 5) = 3x + 3(5)$
$= 3x + 15$

b) $-4(x - 2) = -4x + (-4)(-2)$
$= -4x + 8$

c) $2(3x + 5) = 2(3x) + 2(5)$
$= 6x + 10$

Note: A common student error is to fail to distribute over the second term in parentheses.

Summary:

The distributive property can be extended as follows:

a(b + c + d + ... + n) = ab + ac + ad + ... + an

Example 7: Use the distributive property to remove parentheses

$$a)\ 2(x - y + 5) = 2x + 2(-y) + 2(5)$$
$$= 2x - 2y + 10$$

$$b)\ -3(x^2 + 4x - 5) = -3(x^2) + (-3)(4x) + (-3)(-5)$$
$$= -3x^2 - 12x + 15$$

Summary:

> When no sign or a plus sign precedes parentheses, the parentheses may be removed without having to change the expression inside the parentheses.
>
> When a negative sign precedes parentheses, the signs of all terms within the parentheses are changed when the parentheses is removed.

Example 8: Remove parentheses.

a) $(x - 3)$ $x - 3$

b) $-(x - 3)$ $-x + 3$

c) $-(x^2 + 5)$ $-x^2 - 5$

d) $-(x^2 + 4x - 3)$ $-x^2 - 4x + 3$

Summary

> **To simplify an expression:**
>
> 1. Use the distributive property to remove any parentheses.
>
> 2. Combine like terms.

Example 9. Simplify the following

a) $5 - (4x + 2) = 5 - 4x - 2$
(use distributive property)
$= -4x + 3$
(combine like terms)

b) $5y + 3(y - 4) = 5y + 3y - 12$
(use distributive property)
$= 8y - 12$
(combine like terms)

c) $-3(x + 3) - 4 = -3x - 9 - 4$
(use distributive property)
$= -3x - 13$
(combine like terms)

d) $4(x + 2y) + 2(x - 3y) = 4x + 8y + 2x - 6y$
(use distributive property)
$= 4x + 2x + 8y - 6y$
(rearrange terms)
$= 6x + 2y$
(combine like terms)

Exercise set 2.1

Combine like terms when possible.

1. $3x - 6x$
2. $x + y + 2z$
3. $3x - 2x - 4x$
4. $-2x + 3y - 7y$
5. $5x - 8 + 4$
6. $\frac{2}{3}x + \frac{3}{4} + \frac{3}{4}x$
7. $0.18x + 1.8x$
8. $3x^2 - 8x + 5x^2 - 6x$

Use the distributive property to remove parentheses.

9. $4(x - 5)$
10. $-3(x + 6)$
11. $6(x^2 - 2x + 1)$
12. $2(x - y - 5)$
13. $-2(-y^2 + 3y - 2)$
14. $\frac{3}{4}(4x - 8)$

Simplify when possible.

15. $-(2x^2 - x - 5)$
16. $2x + 3(x - 5)$
17. $4a - (6a - b)$
18. $7y - 2(y - 3)$
19. $5 + (7 - y) + y$
20. $2(2x + y) - 3(x + 4y)$

Answers for exercise set 2.1

1. $-3x$
2. $x + y + 2z$
3. $-3x$
4. $-2x - 4y$
5. $5x - 4$
6. $\frac{17}{12}x + \frac{3}{4}$
7. $1.98x$
8. $8x^2 - 14x$
9. $4x - 20$
10. $-3x - 18$
11. $6x^2 - 12x + 6$
12. $2x - 2y - 10$
13. $2y^2 - 6y + 4$
14. $3x - 6$
15. $-2x^2 + x + 5$
16. $5x - 15$
17. $-2a + b$
18. $5y + 6$
19. 12
20. $x - 10y$

Section 2.2 **The Addition Property of Equality**

Summary:

> An **equation** is a statement that shows two algebraic are equal.
>
> A **linear equation** in one variable is an equation which can be written in the form:
> $ax + b = c$.
> For real numbers a, b, and c, $a \neq 0$.
>
> A **solution** of an equation is the number or numbers that make the equation true.

Example 1. Consider the equation $3x + 2 = 8$.

a) Determine whether $x = 3$ is a solution

  Substitute 3 for x
  $$3(3) + 2 = 8$$
  $$9 + 2 = 8$$
  $$11 = 8 \quad \text{False}$$

  Therefore 3 is not a solution.

b) Determine whether $x = 2$ is a solution.

  Substitute 2 for x
  $$3(2) + 2 = 8$$
  $$6 + 2 = 8$$
  $$8 = 8 \quad \text{True}$$

  Therefore 2 is a solution to the equation.

Example 2. Consider the equation $4(x - 3) - 5x = -12$.

  Determine whether $x = 0$ is a solution

  Substitute 0 for x
  $$4(0 - 3) - 5(0) = -12$$
  $$4(-3) - 5(0) = -12$$
  $$-12 - 0 = -12$$
  $$-12 = -12 \quad \text{True}$$

  Therefore $x = 0$ is a solution to the equation.

Summary:

> **Addition Property of Equality**
>
> If $a = b$, then $a + b = a + c$ for any real number a, b, and c. Since subtraction is defined in term of addition, the addition property of equality allows us to subtract the same number from both sides of the equation.

Example 3. Solve the following equations:

a) $x - 6 = 4$

$x - 6 + 6 = 4 + 6$ Add 6 to both sides
$x + 0 = 10$
$x = 10$

b) $x + 5 = 3$

$x + 5 + (-5) = 3 + (-5)$ Add (-5) to both sides
$x + 0 = -2$
$x = -2$

For the above equation, it could be done in the following manner:

$x + 5 = 3$

$x + 5 - 5 = 3 - 5$ Subtract 5 from both sides
$x + 0 = -2$
$x = -2$

c) $x - .6 = .2$

$x - .6 + .6 = .2 + .6$ Add .6 to both sides
$x + 0 = .8$
$x = .8$

d) $-3 = y + 6$

$-3 + (-6) = y + 6 + (-6)$ Add (-6) to both sides
$-9 = y + 0$
$-9 = y$

Exercise Set 2.2

Solve each equation and check your solution

1) $5 + x = 8$
2) $-3 = y - 7$
3) $x + 8 = -2$
4) $-6 = -8 + a$
5) $0.2 + c = 0.8$
6) $-0.5 = x + 0.5$
7) $-7 = x - 7$
8) $2 + y = 0$
9) $a - 1.6 = 5.3$
10) $-2.6 = -4.3 + x$

Answers for Exercise Set 2.2

1) 3
2) 4
3) -10
4) 2
5) 0.6
6) -1
7) 0
8) -2
9) 6.9
10) 1.7

Section 2.3   **The Multiplication Property of Equality**

Summary:

> Two numbers are **reciprocals** of each other when their product is 1.

Example 1. Give the reciprocals of the following numbers

a) 5 $\frac{1}{5}$ is reciprocal since $\frac{1}{5} \cdot 5 = 1$

b) -6 $-\frac{1}{6}$ is reciprocal since $-\frac{1}{6} \cdot (-6) = 1$

c) $-\frac{2}{3}$ $-\frac{3}{2}$ is reciprocal since $-\frac{2}{3} \cdot -\frac{3}{2} = 1$

Summary:

> **Multiplication Property of Equality**
>
> If $a = b$, then $a \cdot c = b \cdot c$ for any numbers a, b, and c.
>
> The Multiplication Property can be used to solve equations of the form $ax = b$.

Example 2: Solve the following equations

a) $4x = 20$

   $\frac{1}{4} \cdot 4x = \frac{1}{4} \cdot 20$ Multiplication of both sides of

   $\qquad\qquad\qquad$ equation by $\frac{1}{4}$

   $1 \cdot x = 5$
   $x \quad\;\; = 5$

b) $42 = 3x$

$\frac{1}{3} \cdot 42 = \frac{1}{3} \cdot 3x$ Multiplication of both sides of the equation by $\frac{1}{3}$

$14 = 1 \cdot x$
$14 = x$

c) $\frac{x}{3} = 5$

$\frac{3}{1} \cdot \frac{x}{3} = \frac{3}{1} \cdot 5$ Multiply both sides of the equation by 3, the reciprocal of $\frac{1}{3}$

$\frac{x}{1} = 15$
$x = 15$

d) $-\frac{3}{4}x = 9$

$(-\frac{4}{3})(-\frac{3}{4}x) = (-\frac{4}{3})(9)$ Multiply both sides of equation by -4/3

$1 \cdot x = -12$
$x = -12$

e) $1.44 = \frac{x}{3}$

$3(1.44) = 3 \cdot \frac{x}{3}$ Multiply both sides of equation by 3

$4.32 = 1 \cdot x$
$4.32 = x$

Summary:

> Since division can be defined in terms of multiplication (**a/b means a · 1/b**), the multiplication property also allows us to divide both sides of an equation by the same nonzero numbers.

Example 2. Solve the following equations

a) $\quad 6x = -3$

$\quad \dfrac{6x}{6} = -\dfrac{3}{6} \quad$ Divide both sides of the equation by 6

$\quad 1 \cdot x = -\dfrac{3}{6}$

$\quad x = -\dfrac{1}{2}$

b) $\quad 0.16x = 0.8$

$\quad \dfrac{0.16x}{0.16} = \dfrac{0.8}{0.16} \quad$ Divide both sides by 0.16

$\quad 1 \cdot x = 5$
$\quad x = 5$

Note: A common student error is to obtain the equation $-x = 5$ and believe that the equation is solved. The solution must be of the form x = some number.

For any real number a, a ≠ 0, if $-x = a$, then $x = -a$.
Therefore, if $-x = 3$, then $x = -3$.
If $-x = -6$, then $x = -(-6) = 6$.

Exercise Set 2.3

Solve each equation and check your solution.

1. $\quad 3x = 15 \qquad$ 2. $\quad -4x = 24 \qquad$ 3. $\quad 7x = 28$

4. $\quad \dfrac{x}{3} = 5 \qquad$ 5. $\quad \dfrac{x}{-4} = 6 \qquad$ 6. $\quad -6 = \dfrac{x}{7}$

7. $\frac{3}{4}x = -12$ 8. $-\frac{4}{5}x = 10$ 9. $-7 = -\frac{2}{7}x$

10. $100 = -25x$ 11. $-56 = -14x$ 12. $0.25x = 1$

13. $-x = -(-8)$ 14. $-\frac{1}{3}x = -6$ 15. $-\frac{1}{4}x = \frac{2}{3}$

Answers to Exercise Set 2.3

1. 5   2. -6   3. 4   4. 15   5. -24

6. -42   7. -16   8. $-\frac{25}{2}$ 9. $\frac{49}{2}$ 10. -4

11. 4   12. 4   13. 8   14. 18   15. $-\frac{8}{3}$

Set 2.4 Solving Linear Equations with a Variable on only one Side of the Equation

Summary:

> **To Solve Linear Equations with a Variable on only one Side of the Equation**
>
> 1. Use the distributive property to remove the parentheses.
>
> 2. Combine like terms on the same side of the equal sign.
>
> 3. Use the addition property to obtain an equation with the term containing the variable on one side of the equal and a constant on the other side. This will result in an equation of the form $ax = b$.
>
> 4. Use the multiplication property to isolate the variables. This will give a solution of the form $x = b/a$ or $1 \cdot x = b/a$.
>
> 5. Check the solution in the original equation.

Example 1.   Solve the equation $4x + 3 = 11$

$$4x + 3 = 11$$

(step 3) $4x + 3 - 3 = 11 - 3$ Subtract 3 from both sides
$$4x = 8$$

(step 4) $\dfrac{4x}{4} = \dfrac{8}{4}$ Divide both sides by 4

$$x = 2$$

(step 5)   Check: $4x + 3 = 11$
$4(2) + 3 = 11$
$8 + 3 = 11$
$11 = 11$ True

Example 2. Solve the equation $-8 = -6x - 26$

$$-8 = -6x - 26$$

(step 3) $-8 + 26 = -6x - 26 + 26$ Add 26 to both sides
$$18 = -6x$$

(step 4) $\dfrac{18}{-6} = \dfrac{-6x}{-6}$ Divide both sides by -6

$$-3 = x$$

(step 5) Check: $-8 = -6x - 26$
$-8 = -6(-3) - 26$
$-8 = 18 - 26$
$-8 = -8$ True

Example 3. Solve the equation $7x + 3 - 2x = -22$

$$7x + 3 - 2x = -22$$

(step 2) $5x + 3 = -22$ Combine like terms
(step 3) $5x + 3 - 3 = -22 - 3$ Subtract 3 from both sides
$5x = -25$

(step 4) $\dfrac{5x}{5} = \dfrac{-25}{5}$ Divide both sides by 5

$$x = -5$$

(step 5) Check: $7x + 3 - 2x = -22$
$7(-5) + 3 - 2(-5) = -22$
$-35 + 3 + 10 = -22$
$-35 + 13 = -22$
$-22 = -22$ True

Example 4. Solve the equation $x + 1.53 + 0.16x = 3.85$

$$x + 1.53 + 0.16x = 3.85$$

(step 2) $1.16x + 1.53 = 3.85$ Combine like terms
(step 3) $1.16x + 1.53 - 1.53 = 3.85 - 1.53$ Subtract 1.53 from both sides
$1.16x = 2.32$

(step 4) $\dfrac{1.16x}{1.16} = \dfrac{2.32}{1.16}$ Divide both sides by 1.16

$$x = 2$$

Example 4 continued. (step 5) Check: $x + 1.53 + 0.16x = 3.85$
$2 + 1.53 + 0.16(2) = 3.85$
$2 + 1.53 + 0.32 = 3.85$
$3.85 = 3.85$
True

Example 5. Solve the equation $2(2x - 3) = 14$

$2(2x - 3) = 14$

(step 1) $4x - 6 = 14$ Use distributive property

(step 3) $4x - 6 + 6 = 14 + 6$ Add 6 to both sides
$4x = 20$

(step 4) $\dfrac{4x}{4} = \dfrac{20}{4}$ Divide both sides by 4
$x = 5$

(step 5) Check: $2(2x - 3) = 14$
$2(2(5) - 3) = 14$
$2(10 - 3) = 14$
$2(7) = 14$
$14 = 14$ True

Example 6. Solve the equation $-3(x - 5) + 2x = 5$

$-3(x - 5) + 2x = 5$

(step 1) $-3x + 15 + 2x = 5$ Use distributive property

(step 2) $-x + 15 = 5$ Combine like terms

(step 3) $-x + 15 - 15 = 5 - 15$ Subtract 15 from both sides
$-x = -10$

(step 4) $\dfrac{-x}{-1} = \dfrac{-10}{-1}$ Divide both sides by -1
$x = 10$

(step 5) Check: $-3(x - 5) + 2x = 5$
$-3(10 - 5) + 2(10) = 5$
$-3(5) + 2(10) = 5$
$-15 + 20 = 5$
$5 = 5$ True

Exercises Set 2.4

Solve the following equations.

1. $4x - 6 = 10$
2. $3x + 7 = 10$
3. $-12 = 6x + 12$
4. $2x - 3 + 5x = 11$
5. $-2 = -2x - 6 - 2x$
6. $10x + 3 - 4x = 7 - 16$
7. $-2x + 3 = -9$
8. $-\frac{3}{4}x - 5 = 4$
9. $0.6 + 0.6x = -1.8$
10. $x - 2x = 6$
11. $-7 = -(x - 4)$
12. $4(x - 3) = 16$
13. $x - (4x + 5) = -11$
14. $10 = 2(x = 5) - 6x$
15. $4(x + 3) + 2(2x - 6) = 32$
16. $3(2x - 5) + 2x = -23$
17. $x - 0.35x = 1.95$
18. $0.1(2x - 3) = 0$

Answers for Exercises Set 2.4

1) 4   2) 1   3) -4   4) 2
5) -1   6) -2   7) 6   8) -12
9) -4   10) -6   11) 11   12) 7
13) 2   14) 0   15) 4   16) -1
17) 3   18) $\frac{3}{2}$ or 1.5

Section 2.5   Solving Linear Equations with the Variables on Both Sides of the Equation

Summary:

> **To Solve Linear Equations with the Variable on Both Sides of the Equal Sign**
>
> 1. Use the distributive property to remove parentheses
>
> 2. Combine like terms on the same side of the equal sign.
>
> 3. Use the addition property to rewrite the equation with all terms containing the variable on one side of the equal sign and all terms not containing the variable on the other side of the equal sign. It may be necessary to use the addition property to accomplish this goal. You will eventually get an equation of the form $ax = b$.
>
> 4. Use the multiplication property to isolate the variable. This will give a solution of the form $x =$ some number.
>
> 5. Check the solution to the original equation.

Example 1   Solve the equation $7y - 4 = 4y + 20$

$$7y - 4 = 4y + 20$$

Steps 1 and 2 are not needed

(step 3) $7y - 4 - 4y = 4y - 4y + 20$ Subtract 4y from both sides
$$3y - 4 = 20$$

(step 3) $3y - 4 + 4 = 20 + 4$ Add 4 to both sides
$$3y = 24$$

54

(step 4) $\dfrac{3y}{3} = \dfrac{24}{3}$ Divide both sides of equation by 3

           $y = 8$

(step 5)   Check: $7y - 4 = 4y + 20$
                   $7(8) - 4 = 4(8) + 20$
                   $56 - 4 = 32 + 20$
                   $52 = 52$ True

Example 2. Solve the equation $4r - 5r - 15 = 2r$

           $4r - 5r - 15 = 2r$ Step 1 not needed
(step 2) $-r - 15 = 2r$ Combine like terms
(step 3) $-r + r - 15 = r + 2r$ Add r to both sides
           $-15 = 3r$
(step 4) $-\dfrac{15}{3} = \dfrac{3r}{3}$ Divide both sides by 3
           $-5 = r$

(step 5)   Check: $4r - 5r - 15 = 2r$
                   $4(-5) - 5(-5) - 15 = 2(-5)$
                   $-20 + 25 - 15 = -10$
                   $-10 = -10$ True

Example 3. Solve the equation $3 - v + 7 = -1 - 2(2 - v)$

(step 1) $3 - v + 7 = -1 - 4 + 2v$ Distributive property was used
(step 2) $10 - v = -5 + 2v$ Combine like terms
(step 3) $10 - 10 - v = -5 - 10 + 2v$
           $-v = -15 + 2v$
(step 3) $-v - 2v = -15 + 2v - 2v$ Subtract $2v$ from both sides
           $-3v = -15$
(step 4) $\dfrac{-3v}{-3} = \dfrac{-15}{-3}$ Divide both sides by -3
           $v = 5$

(step 5)   Check: $3 - v + 7 = -1 - 2(2 - v)$
                   $3 - 5 + 7 = -1 - 2(2 - 5)$
                   $5 = -1 - 2(-3)$
                   $5 = -1 + 6$
                   $5 = 5$ True

Example 4. Solve the equation $0.1(x - 20) = 0.4x + 1$

	$0.1(x - 20)$	$= 0.4x + 1$	
(step 1)	$0.1x - 2$	$= 0.4x + 1$	Distributive property was used Step 2 not needed
(step 3)	$0.1x - 0.4x - 2 = 0.4x - 0.4x + 1$		Subtract $0.4x$ from both sides
	$-0.3x - 2 = 1$		
(step 3)	$-0.3x - 2 + 2$	$= 1 + 2$	Add 2 to both sides
	$-0.3x$	$= 3$	
(step 4)	$\dfrac{-0.3x}{-0.3}$	$= \dfrac{3}{-0.3}$	Divide both sides by -0.3
	$x = -10$		

(step 5) Check:
$$0.1(x - 20) = 0.4x + 1$$
$$0.1(-10 - 20) = (0.4)(-10) + 1$$
$$0.1(-30) = -4 + 1$$
$$-3 = -4 + 1$$
$$-3 = -3 \quad \text{True}$$

Summary:

> Equations that are true for all values of x are called **identities**.
>
> Equations for which you obtain an obviously false statement have no solution.

Example 5. Solve the equation $3(x - 2) = 6x - 3x - 6$

	$3(x - 2)$	$= 6x - 3x - 6$	
(step 1)	$3x - 6$	$= 6x - 3x - 6$	The distributive property was used
(step 2)	$3x - 6$	$= 3x - 6$	Combine like terms
(step 3)	$3x - 3x - 6 = 3x - 3x - 6$		Subtract $3x$ from both sides
	-6	$= -6$	
(step 3)	$-6 + 6$	$= -6 + 6$	Add 6 to both sides
	0	$= 0$	

Therefore, since 0 = 0 is always true, the equation is true for all value of x.

Example 6. Solve the equation $2x + 19 - x = 8x + 3 - 7x$

$$2x + 19 - x = 8x + 3 - 7x$$ Step 1 is not needed

(step 2) $x + 19 = x + 3$ Combine like terms

(step 3) $x - x + 19 = x - x + 3$ Subtract x from both sides
        $19 = 3$

Since 19 = 3 is false, the equation has no solution.

Exercises Set 2.5

Solve each equation.

1. $6y + 2 = y + 17$
2. $11n + 3 = 10n + 11$
3. $3x + 1 = 11$
4. $4x - 7 = 5x + 1$
5. $5a + 7 = 2a + 7$
6. $6a + 3 + 2a = 11$
7. $5y + 9 + 2y = 30 - 7$
8. $5x + 2(x + 1) = 23$
9. $6y + 2(2y + 3) = 16$
10. $\frac{1}{2}(4x + 10) = -\frac{1}{3}(9 - 12 - x)$
11. $4 - 3a = 7 - 2(2a + 5)$
12. $9 - 5x = 12 - (6x + 7)$
13. $3x + 4 + 7x = 2(5x + 2)$
14. $5x + 9 = 5x - 15$

Answers to Exercises Set 2.5:

1. 3
2. 8
3. 2
4. -8
5. 0
6. 1
7. 2
8. 3
9. 1
10. 4
11. -7
12. -4
13. all real numbers
14. no solution

Section 2.6   **Ratios and Proportions**

Summary:

> A **ratio** is a quotient of two quantities. The ratio of **a** to **b** may be written as **a:b** or **a/b**.
>
> The **a** and **b** are called **terms** of the ratio.

Example 1. A class consists of 14 males and 17 females.

a) Find the ratio of females to males   17:14 or $\frac{17}{14}$

b) Find the ratio of males to females   14:17 or $\frac{14}{17}$

c) Find the ratio of males to the entire class   14:31 or $\frac{14}{31}$

Example 2. Find the ratio of 8 inches to 2 feet

We must first express with the same unit. So 2 feet will be expressed as 2(12 inches) = 24 inches.

Thus 8:24 or $\frac{8}{24}$ or $\frac{1}{3}$

Summary:

A **proportion** is a special type of equation. It is a statement of equality between 2 ratios. One way of denoting a proportion is a:b = c:d which is read "a is to b as c is to d." In this text we write proportions as
   a/b = c/d

The a and d are referred to as **extremes**, and the b and c are referred to as **means** of the proportion. One method that can be used in evaluating proportions is **cross-multiplication**.

Cross-Multiplication

If a/b = c/d, then ad = bc.

Example 3. Solve the following for x using cross multiplication

a) $\dfrac{x}{8} = \dfrac{9}{4}$ Check: $\dfrac{x}{8} = \dfrac{9}{4}$

   $4 \cdot x = 8 \cdot 9$ $\dfrac{18}{8} = \dfrac{9}{4}$

   $4x = 72$ $\dfrac{9}{4} = \dfrac{9}{4}$ True

   $\dfrac{4x}{4} = \dfrac{72}{4}$

   $x = 18$

b) $\dfrac{3}{-x} = \dfrac{4}{16}$ Check: $\dfrac{3}{-x} = \dfrac{4}{16}$

$\quad\quad 3 \cdot 16 = -x \cdot 4$ $\quad\quad\quad\quad\quad\quad \dfrac{3}{-(-12)} = \dfrac{4}{16}$

$\quad\quad 48 = -4x$ $\quad\quad\quad\quad\quad\quad\quad\quad \dfrac{3}{12} = \dfrac{4}{16}$

$\quad\quad \dfrac{48}{-4} = \dfrac{-4x}{-4x}$ $\quad\quad\quad\quad\quad\quad\quad \dfrac{1}{4} = \dfrac{1}{4}$ True

$\quad\quad -12 = x$

Summary:

> **To Solve Problems Using Proportions**
>
> 1. Represent the unknown quantity by a letter.
>
> 2. Set up the proportion by listing the given ratio on the left side of the equal sign and the unknown and the other quantity on the right side of the equal sign. When setting up the right side of the proportion, the same respective quantities should occupy the same respective portions on the left and right. For example, an acceptable proportion might be
>
> $\quad$ Given ratio $\left\{\dfrac{\text{miles}}{\text{hours}} = \dfrac{\text{miles}}{\text{hours}}\right.$
>
> 3. Once the proportion is correctly written, drop the units and cross-multiply.
>
> 4. Solve the resulting equation.
>
> 5. Answer the questions asked.

Note: The two ratios must have the same units. For example, if one ratio is given in miles/hour and the other ratio is given in feet/hour, one of the ratios must be changed before setting up the proportion.

Example 4:

During the first 560 miles of their vacation trip, the Smith's auto used 18 gallons of gasoline. At this rate, how many gallons to the nearest tenth of a gallon will be needed to finish the remaining 420 miles of the trip?

(step 1)   Let x = number of gallons to finish the trip

(step 2)   Given ratio is 560 miles to 18 gallons
Ratio with unknown is 420 mile to x gallons
Proportion is

$$\frac{560 \text{ miles}}{18 \text{ gallons}} = \frac{420 \text{ miles}}{x \text{ gallons}}$$

(step 3) $\frac{560}{18} = \frac{420}{x}$

560x = 18 · 420
560x = 7560

(step 4) $\frac{560x}{560} = \frac{7560}{560}$

x = 13.5

(step 5)   13.5 gallons are needed to complete the remaining 420 miles of the trip.

Example 5.

A recipe calls for 2 lbs of beef to prepare 12 servings. If we have available 12 ozs of beef, how many servings will that make?

(step 1)   Let x = number of servings with 12 ozs available

(step 2)   First convert 2 lbs to ozs
2lbs = 2(16ozs) = 32 ozs
Given ratio is 32 ozs to 12 servings
Ratio with unknown is 12 ozs to x servings

$$\frac{32 \text{ ozs}}{12 \text{ servings}} = \frac{12 \text{ ozs}}{x \text{ servings}}$$

(step 3) $\dfrac{32}{12} = \dfrac{12}{x}$

$32x = 12 \cdot 12$
$32x = 144$

(step 4) $\dfrac{32x}{32} = \dfrac{144}{32}$

$x = 4.5$

(step 5) 12 ounces of beef will make 4.5 servings

Summary:

> Proportions can be used to convert from one quantity to another. For example, you can use proportion to convert a measurement in feet to a measurement in meters, or to convert from U.S. dollars to Mexican pesos.

Example 6.

300 dollars to British pounds if the exchange rate is $1.61 equals one British pound. (Convert)

(step 1) Let x = number of British pounds equal to 300 dollars

(step 2) Given ratio is 1 British pound to 1.61 dollars
Ratio with unknown quantity is x British pounds to 300 dollars

$$\dfrac{1.61 \; dollars}{1 \; British \; pound} = \dfrac{300 \; dollars}{x \; British \; pounds}$$

(step 3) $\dfrac{1.61}{1} = \dfrac{300}{x}$

$1.61x = 300 \cdot 1$
$1.61x = 300$

(step 4) $\dfrac{1.61x}{1.61} = \dfrac{300}{1.61}$

$x = 186.34$

(step 5)   300 dollars is equivalent to 186.34 British pounds

Summary:

> Proportions can be used to solve problems in geometry and trigonometry. If two figures are **similar**, then their corresponding sides are in proportion.

Example 7. The two triangles below are similar.

The following measurements are known AB = 8 inches, XY = 5 inches, and AC = 5 inches. Find XZ.

(step 1)   Let x = Length of XZ

(step 2)   Known ratio is AB to XY which is 8 to 5
            Ratio with unknown quantity is AC to XZ which is 5 to x

$$\frac{8 \text{ inches}}{5 \text{ inches}} = \frac{5 \text{ inches}}{x \text{ inches}}$$

(step 3) $\frac{8}{5} = \frac{5}{x}$

            $8x = 5 \cdot 5$
            $8x = 25$

(step 4) $\frac{8x}{8} = \frac{25}{8}$

            $x = 3\frac{1}{8}$

(step 5)   The length of XZ is $3\frac{1}{8}$ inch

Exercises Set 2.6

If the distribution of grades for a particular exam is 5 A's, 6 B's, 10 C's, 6 D's, and 4 F's, write the ratios of the following:

1. A's to F's
2. C's to B's
3. Total grade to B's

Determine the following ratio. Express each ratio in lowest terms:

4. 12 yards to 20 yards
5. 60 minutes to 45 minutes
6. 80 minutes to 2 hours
7. 25 feet to 5 yards

Solve for the variable by cross-multiplying:

8. $\dfrac{3}{x} = \dfrac{12}{40}$
9. $\dfrac{x}{11} = \dfrac{40}{10}$
10. $\dfrac{15}{48} = \dfrac{-x}{16}$
11. $\dfrac{15}{4} = \dfrac{30}{-x}$

Write a proportion that can be used to solve the problem then solve the problem.

12. If 100 lbs of pesticide will cover 50 acres of potatoes, how much pesticide is needed for 135 acres of potatoes?

13. A homeowner in Florida pays $360 in property taxes on his home, which is assessed at $80000. His neighbor's house is assessed at $145,000. How much property tax will his neighbor pay?

14. A recipe calls for 3 lbs of flour to make a cake that will serve 16 persons. If only 20 ozs of flour is available to make a cake, how many persons will that cake serve?

15. A meter is equal to 39.37 inches. How many meters is equivalent to 90 inches? (Round to the nearest tenth of a meter)

16. Find the unknown length of the two similar triangles below: (Triangle ABC is similar to XYZ)

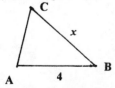
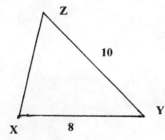

17. Find the unknown length of the two similar triangles below:
(Triangle DEF is similar to triangle UVW)

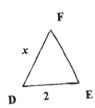 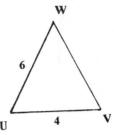

Answers to Exercises Set 2.6:

1. 5:4   2. 10:6 or 5:3   3. 31:6

4. 3:5   5. 4:3   6. 2:3

7. 5:3   8. 10   9. 44

10. -5   11. -8   12. 270 lbs

13. $652.50   14. $6\frac{2}{3}$ persons   15. 2.3 meters

16. 5   17. 3

Section 2.7   **Inequalities in one variable**

Summary:

> The greater-than symbol, $>$, and the less-than symbol, $<$, were introduced in section 1.4. The symbol $\geq$ means greater than or equal to and $\leq$ means less than or equal to. A mathematical statement containing one or more of these symbols is called an **inequality**. The direction of the symbol is called the **sense** of the inequality.
>
> **Examples of Inequalities in One Variable**
>
> $x + 3 < 5$, $x + 4 \geq 2x - 6$,
>
> $4 > -x + 3$
>
> To solve an inequality, we must get the variable by itself on one side of the inequality symbol.

We make use of properties very similar to those used to solve equations.

Summary

> **Properties Used to Solve Inequalities**
>
> For real numbers a, b, and c:
>
> 1. If a > b, then
>   a + b > b + c
>
> 2. If a > b, then
>   a - c > b - c
>
> 3. If a > b **and c > 0**, then
>   ac > bc
>
> 4. If a > b **and c > 0**, then
>   a/c > b/c

Example 1.

Solve the inequality x + 3 > 7, and graph the solution on the real number line.

```
x + 3       > 7
x + 3 - 3   > 7 - 3    Subtract 3 from both sides of
                              the inequalities
x           > 4
```

<----------|-|-|-|-○------------>
           0   4

The solution is all real numbers greater than 4. Note that the open circle indicates that 4 is not part of the solution.

Example 2.

Solve the inequality 3x + 6 ≤ 18, and graph the solution on the real number line.

```
3x + 6        ≤ 18
3x + 6 -6     ≤ 18 - 6   Subtract 6 from both sides of the
                            inequality
3x            ≤ 12

3x            ≤ 12       Divide both sides of the
3                3          inequality by 3
x             ≤ 4
```

The solution is all real numbers less than or equal to 4.

<------------------------------------->

The closed circle at 4 indicates that 4 is part of the solution.

Summary:

> When an inequality is multiplied or divided by a negative number the direction of the inequality symbol changes.
>
> <u>Additional Properties Used to Solve Inequalities:</u>
>
> 5. If $a > b$ <u>and $c < 0$</u>, then $ac < bc$
>
> 6. If $a > b$ <u>and $c < 0$</u>, then $a/c < b/c$

Example 3.

Solve the inequality $-4x + 3 \leq -9$, and graph the solution on the real number line.

$-4x + 3 \leq -9$

$-4x + 3 - 3 \leq -9 - 3$ Subtract 3 from both sides of the inequality

$-4x \leq -12$

                                     Divide both sides of the inequality by -4 and change the direction of the inequality symbol.

$x \geq 3$

The solution is all real number greater than or equal to 3.

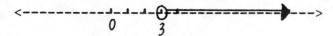

Example 4.

Solve the inequality $6x - 11 < 8x + 7$, and graph the solution on the real number line.

$6x - 11$	$< 8x + 7$	
$6x - 8x - 11$	$< 8x - 8x - 7$	Subtract 8x from both sides of the inequality.
$-2x - 11$	< -7	
$-2x - 11 + 11$	$< -7 + 11$	Add 11 to both sides of the inequality
$-2x$	< 4	
$\dfrac{-2x}{-2}$	$> \dfrac{4}{-2}$	Divide both sides by -2 and change the direction of the inequality.
x	> -2	

<-------------○--┼--┼-------------►------>
                -2   0

Example 5.

Solve the inequality $-3(x + 4) \geq -12$, and graph the solution on the real number line.

$-3(x + 4)$	≥ -12	
$-3x - 12$	≥ -12	Use of distributive property
$-3x - 12 + 12$	$\geq -12 + 12$	Add 12 to both sides of the inequality
$-3x$	≥ 0	
$\dfrac{-3x}{-3}$	$\leq \dfrac{0}{-3}$	Divide by -3 ; Change sign of inequality
x	≤ 0	

<--◄-------------●------------->
                 0

Note: If in solving an inequality you arrive at a statement that is always true, such as $4 < 8$, then the solution is all real numbers. If you arrive at a statement that is false, such as $2 \geq 3$, then there is no solution.

Example 6.

Solve the inequality $2(x-2) \geq 4x - 2x$ and graph the solution on the real number line.

$$2(x - 2) \geq 4x - 2x + 5$$

$$2x - 4 \geq 4x - 2x + 5 \quad \text{Use the distributive property}$$

$$2x - 4 \geq 2x + 5 \quad \text{Combine like terms}$$

$$2x - 2x - 4 \geq 2x - 2x + 5 \quad \text{Subtract 2x from both sides of the inequality}$$

$$-4 \geq 5 \quad \text{False}$$

Therefore, no solution to the inequality.

Exercises Set 2.7

Solve the inequalities, and graph the solution on the real number line.

1. $x - 5 < 12$ 2. $x + 3 \leq 6$

3. $-5x \leq 25$ 4. $21 \leq 7x$

5. $\frac{3}{4}x > 6$ 6. $16 \geq -2x$

7. $10 - x > 4$ 8. $3x + 5 \leq 14$

9. $-3x - 7 \leq 8$ 10. $\frac{x}{5} + 3 \leq -1$

11. $3(2x - 5) \geq 8x - 5$ 12. $5x - 8 \geq 7x - 9$

13. $2(2y - 5) \leq 3(5 - 2y)$ 14. $2(5x - 8) \leq 7(x - 3)$

15. $5(2 - x) > 3(2x - 5)$ 16. $4(3d - 1) > 3(2 - 5d)$

17. $5(x - 2) > 9x - 3(2x - 4)$

18. $15 - 5(3 - 2x) \leq 4(x - 3)$

19. $2x + 3x - 6 < 5x + 2$ 20. $4(x + 3) \geq 3x + x + 15$

Answers to Exercises Set 2.7:

1. $x < 17$
2. $x \leq 3$
3. $x \geq -5$
4. $x \geq 3$
5. $x > 8$
6. $x \geq -8$
7. $x < 6$
8. $x \leq 3$
9. $x \geq -5$
10. $x \leq -20$
11. $x \leq -5$
12. $x \leq 1/2$
13. $y \leq 5/2$
14. $x \leq -5/3$
15. $x < \dfrac{25}{11}$
16. $d > \dfrac{10}{27}$
17. $x > 11$
18. $x \leq -2$
19. all real numbers
20. no solution

Practice Test Chapter 2:

Simplify

1. $-2(3x + 5)$
2. $-(x - y + 5)$

Simplify when possible.

3. $3x - 5 + 7x - 6$
4. $3a - 7a + 4b$
5. $2(x - 3) + 3x + 2$
6. $-(x - 3) - x + 3$

Solve each equation:

7. $3x + 5 = 11$
8. $\frac{3}{4} x = -9$
9. $3x - 2 = 5x + 8$
10. $3(2x - 5) = 8x - 9$
11. $11 - 4x = 2x + 8$
12. $6 - 2(5x - 8) = 3(3x - 4)$
13. $6x - 3(2 - 3x) = 4(2x - 7)$
14. $5x - 6 = 5(x - 1) - 1$
15. $\frac{-x}{7} = \frac{25}{10}$

Solve each inequality and graph the solution on a real number line.

16. $2x - 7 \leq 13$
17. $-2(3x + 2) > 3x + 5$
18. $3(x - 2) + 5x < 8x$

19. The following triangles are similar. ABC is similar to XYZ. Find the length of the unknown side.

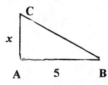
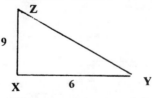

20. If a car travels 240 miles on 11 gallons of gasoline, how far would it travel on 7 gallons of gasoline ?. (Round to the nearest tenth of a mile.)

Answers to Practice Test Chapter 2

1. $-6x - 10$
2. $-x + y - 5$
3. $10x - 11$
4. $-4a + 4b$
5. $5x - 4$
6. $-2x + 6$
7. 2
8. -12
9. -5
10. -3
11. $1/2$
12. 2
13. $-22/7$
14. all real numbers
15. -17.5
16. $x \leq 10$
17. $x < -1$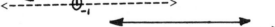
18. all real numbers

19. $7\frac{1}{2}$ or $15/2$

20. 152.7 miles

Section 3.1   **Formulas**

Summary:

> The **simple interest** formula is interest = principal x rate x time or i = prt.

Example 1   Stan borrows $1000 from a bank for 2 years at a rate of 14% simple interest per year. How much will Stan owe the bank after 2 years?

Solution: Use i = prt

Here p = 1000, i = 14% or .14 and t = 2.
So i = 1000(.14)(2) = $280
The total amount he owes the bank is $1000 + $280 or $1280.

Summary:

> 1. The **perimeter** of a figure is the sum of the lengths of the sides of a figure and is measured in units such as inches, centimeters, feet, and so on.
>
> 2. The **area** of a figure is the number of square units contained inside the figure and is measured in cm^2, in^2, ft^2, and so on.
>
> 3. The table in your text book lists all the formulas for perimeter and area of common geometric figures such as squares, rectangles, parallelograms, trapezoids, triangles, and circles.

Example 2   Find the total length of base moulding needed along the base of the floor walls of a bedroom which measures 16'x10'.

Solution.   Assume the bedroom is rectangular
perimeter = 2l + 2w
p = 2(16) + 2(10)
p = 32 + 20 = 52 feet

Example 3.   The perimeter of a rectangle is 48cm. If the width is 8", find the length.

Solution.   p = 2l + 2w
48 = 2l + 2(8)
48 = 2l + 16
48 - 16 = 2l + 16 - 16

$$\frac{32}{2} = \frac{2l}{2}$$

16 = l
Length is 16"

Example 4   A triangle has an area of 64 ft² and a height of 8 ft. What is the length of the base of this triangle?

Solution.

$$A = \frac{1}{2} bh$$

$$64 \text{ ft}^2 = \frac{1}{2} (b)(8)$$

$$64 = \frac{1}{2} (8)b$$

$$\frac{64}{4} = \frac{4b}{4}$$

16 = b
So the base is 16 ft.

Summary:

> 1. The **circumference**, c, of a circle is the perimeter of the curve that forms the circle.
>
> 2. The **radius**, r, of a circle is the line segment from the center of a circle to any point on the circle.
>
> 3. A **diameter** of a circle is a line segment through the center whose endpoints both lie on the circle.

Example 5.   The diameter of a pizza pie is 14". What is the area and circumference of the pizza.

Solution.

Since $A = \pi r^2$ and $r = \frac{1}{2} d$,

$r = \frac{1}{2}(14) = 7"$

So $A = \pi(7)^2$ or $3.14 \cdot (7)^2 = 3.14(49) = 153.86$ square inches.

$C = 2\pi r$ or $C = \pi d$
We can use either formula
Using $C = 2\pi r$ we have $C = 2(3.14)(7) = 14 \cdot (3.14) = 43.96$ in. (Note the units)

Or $C = \pi d = 3.14(14) = 43.96$ inches

Example 6   The volume of a right circular cylinder is 706.5 in³. What is the radius of the cylinder if the height is 9"?

Solution.

Use $V = \pi r^2 h$
$706.5 = (3.14)r^2(9)$

$$\frac{706.5}{28.26} = \frac{28.26 r^2}{28.26}$$

Example 6 continued

$$25 = r^2$$

If $r^2 = 25$ then $r = \sqrt{25}$ or $r = 5"$

Example 7.   Solve $V = l \cdot w \cdot h$ for w.

Solution.
1. Treat all letters as constant <u>except</u> for w.
2. Solve for w by isolating it on one side of the equation.

$$V = l \cdot w \cdot h$$

$$\frac{V}{lh} = \frac{l \cdot h \cdot w}{lh} \qquad \text{rearrange terms}$$

$$\frac{V}{lh} = w \qquad \text{isolate w}$$

Example 8.   Solve $F = \frac{9}{5} c + 32$ for c.

Solution.

$$5F = 5\left(\frac{9}{5} c + 32\right)$$

$$5F = 5\left(\frac{9}{5} c\right) + 5(32)$$

$$5F - 160 = 9c + 160 - 160$$

$$\frac{5F - 160}{9} = \frac{9c}{9}$$

$$\frac{5}{9} F - \frac{160}{9} = c$$

Example 9   Solve $A = \frac{m + d}{2}$ for m.

Solution.

$$2A = \frac{m+d}{2} \cdot \frac{2}{1}$$

$$-d + 2A = m + d - d$$

$$2A - d = m$$

Exercise Set 3.1

Use the formula to find the value of the indicated variable. Round answers to nearest hundredth.

1. $A = \frac{1}{2}h(b + c)$. Find A when $h = 6$, $b = 3$, $c = 5$.

2. $P = 2l + 2w$; Find l when $P = 40$ and $w = 16$.

3. $Z = \frac{x - a}{5}$; Find z when $x = 130$, $a = 100$ and $s = 15$.

Solve each equation for y. Then find the value of y for the given value of x:

4. $3x - 2y = 6$, when $x = 3$

5. $2x - y = 10$, when $x = -4$

6. $3x - 4y = 12$, when $x = 5$

Solve for the variable indicated:

7. $I = prt$ for r

8. $ax + by = c$ for x

9. $V = \frac{1}{3} \pi r^2 h$ for h

10. $A = \frac{x + y + z}{4}$ for z

11. $P = 2l + 2w$ for l

12. $V = l \cdot w \cdot h$ for w

13. $Z = \dfrac{x - m}{5}$ for m

14. If $C = \dfrac{5}{9}(F - 32)$, find C if F = 95°

15. Joe left $2000 in a bank which paid $4\frac{1}{2}$% simple interest. When he withdrew his money, he received $360.00 of interest. How long did he leave his money in the account?

16. A vacant lot is in the shape of a trapezoid. If the height of the trapezoid is 12 meters, one base is 15 meters and the area is 126 square meters, find the length of the other base.

17. If the circumference of a car tire is 87.92 inches, how large is the diameter of the tire?

18. If a soup can measures 3 inches in diameter and is 7" high, what is its volume?

19. Find the area of a triangle whose base is 5" and whose height is 9".

20. If the sum, S, of the first n even numbers is $S = n^2 + n$, find the sum of the first 20 even numbers.

Answers to exercise set 3.1

1. 24

2. 1 = 4

3. 2

4. $y = \dfrac{3}{2}x - 3$, $1\dfrac{1}{2}$

5. $2x - 10 = y$, $-18 = y$

6. $\dfrac{3}{4}x - 3 = y$, $-\dfrac{3}{4}$

7. $\dfrac{I}{pt} = r$

8. $x = -\dfrac{by}{a} + \dfrac{c}{a}$

9. $\dfrac{3V}{\pi r^2} = h$

10. $4A - x - y = z$

11. $\dfrac{P - 2w}{2} = l$

12. $\dfrac{V}{lh} = w$

13. $m = x - 5z$

14. $C = 35°$

15. 4 years

16. $6m = b$

17. 28 inches

18. $V = 197.82 \text{ in}^3$

19. 22.5 in^2

20. $S = 420$

Section 3.2   Changing Application Problems into Equations

Summary:

> Statements can be represented as algebraic expressions.

The following verbal statements have been translated to algebraic statements.

Verbal	Algebraic
a number increased by 6	$x + 6$
8 more than a number	$x + 8$
5 less than a number	$x - 5$
a number decreased by 7	$x - 7$
3 times a number	$3x$
four fifths of a number	$\frac{4}{5}x$
a number divided by 4	$\frac{x}{4}$
7 more than 3 times a number	$3x + 7$
6 less than 4 times a number	$4x - 6$
2 times the difference of a number and 5	$2(x - 5)$

Example 1   Express each phrase as an algebraic expression:

1. The area increased by 8 cm²

    $A + 8$

2. 5 more than twice the perimeter

    $2p + 5$

3. twice the sum of the length and 5

   $2(l + 5)$

4. 5 more than twice the volume

   $5 + 2V$ or $2V + 5$

Example 2   Write three different verbal statements to represent the following expression.

$7x + 3$

Solution: 1) Three more than seven times a number
            2) Seven times a number increased by 3
            3) The sum of seven times a number and 3

Note: Some students may incorrectly interpret this last verbal statement as $7(x + 3)$. However, this mathematical statement should be interpreted as the product of 7 and the sum of a number and 3.

Example 3   Write a verbal statement to represent each expression. There can be more than one answer.

a) $5x - 2$

   Two less than 5 times a number

b) $5(x - 2)$

   Five times the difference of a number and 2

Example 4   Represent the following statement mathematically:

The number of violent crimes decreased by 9%

Solution.

Let c = previous number of violent crimes
Then $c - .09c$ represents the present number of violent crimes.

This statement is of the general form:
Original amount decreased by a percentage of the original amount equals the new amount. Or $A - r\%A = $ new A.

Example 5   The population of a city increased by 4%. Represent this statement mathematically.

Solution.

$p + 4\%p = \text{new } p$
$p + .04p = \text{new } p$

Common error: Some students write the answer to this question as $p + .04$. It is important to realize that a percent of a quantity must always be a percent multiplied by some letter or number.

Example 6   A 25 foot is cut into 2 pieces. If x represents one piece, then express the length of the other piece in terms of x.

Solution.

Notice the pattern which follows:

1st piece	2nd piece
5	20 = 25 - 5
8	17 = 25 - 8
x	25 - x

Example 7   A 15 foot board is cut into 2 pieces. One piece is twice as long as the other piece. Find an algebraic expression for the length of the 2 pieces.

Solution.

        x   2x

Note: We also know that $x + 2x = 15$

Example 8   Express each of the following as an algebraic expression:

a) Cost of purchasing y items at $5 each

   5y

b) 15% commission on x dollars in sales

   .15x

c) Cost of producing y items at 35¢ each and x items at 54¢ each.

   .35y + .54x

d) Distance traveled in t hours at a rate of 65 miles per hour.

   65t

e) The numbers of ml of intravenous fluid in x hours at a rate of 150 ml of fluid per hour.

   150x

f) The number of seconds in a minutes and b seconds.

   $60a + b$

g) The sum of 2 consecutive integers if the first integer is n.

   Example: $3 + 4$
   $5 + 6$
   $7 + 8$
   $100 + 101$ so
   $(n) + (n + 1)$

h) The product of 2 consecutive **odd** integers if the first **odd** integer is x.

   example: $3 \cdot 5$
   $5 \cdot 7$
   $101 \cdot 103$
   $x \cdot (x + 2)$

Summary:

> The word "is" in a verbal problem means "is equal to" and is represented by an equal sign.

Example 9 The sum of three times a number and four **is** 7.

Equation: $3x + 4 = 7$

Example 10 The product of 2 consecutive even integers is 80.

$n(n + 2) = 80$

Example 11   A number increased by 20% is 12.

$n + .20n = 12$

Example 12   Write a verbal statement for $y + 3(y - 5) = 7$.

The sum of a number y and 3 times the difference of y and 5 is seven.

Example 13   The cost of a skirt at a 25% off sale is $65.

$x - 0.25x = 65$

Example 14   Write an algebraic expression for

"One car travels 5 times as far as a second car. The total distance traveled is 720 miles."

$5x + x = 720$

Exercise Set 3.2

Write as an algebraic expression

1. 5 times a number

2. 7 less than twice a number

3. 18% of a number, y

4. $8\frac{1}{2}$ % interest rate on p dollars

5. Five less than 4 times a number

6. Four times the sum of a number and 9

7. The cost of renting y video cassetes at a rate of $2.25 per video casseste.

Express as a verbal statement.

8. $5 - y$ 9. $3x + 2$ 10. $4(x - 5)$

Select a variable to represent one quantity. State what that variable represents. Then express the second quantity in terms of the variable.

11. John's salary is $150 more than Harry's Salary.

12. The cost of an item and the cost increased by a $6\frac{1}{2}$% sales tax.

13. The federal deficit and the federal deficit reduced by 15%.

14. The world population and the world population increased by 7%.

Express as an equation.

15. One board is 3 times as long as a second board. The sum of their lengths is 20

16. The cost of a stereo plus 7% tax is $800.

17. One train travels 5 miles less than 3 times another train. The total distance traveled by both trains is 500 miles.

18. The product of 2 consecutive integers is 132.

19. The sum of 2 consecutive even integers is 70.

20. The total cost of a company is $500 per month plus $35 for each item produced. The total cost this month is $1200.

21. The sum of twice a number and that number increased by 3 is 8.

22. The cost of traveling x miles at 29 cents per mile is $24.90.

Answers to exercise set 3.2

1. 5x   2. 2x − 7   3. .18y   4. .085p   5. 4x − 5

6. 4(x + 9)   7. 2.25y   8. The number 5 decreased by y

9. Three times a number increased by 2.

10. The product of 4 and 5 less than a number.

11. x = Harry's salary; 150 + x = John's salary

12. x = cost of an item ; x + .065x = cost of the item plus sales tax

13. d = federal deficit; d − .15d

14. p = world's population; p + .07p

15. x = length of short board; x + 3x = 20

16. x = cost of stereo before tax. x + .07x = 800

17. d = distance traveled by one train; 3d - 5 = distance traveled by the second train; d + (3d - 5) = 500

18. n = first integer; n(n + 1) = 132

19. n = first even integer; n + n + 2 = 70

20. x = number of items; 500 + 35x = 1200

21. x = a number; 2x + (x + 3) = 8

22. x = number of miles driven; .29x = 24.90

Section 3.3   **Solving Application Problems**

1. **Summary**

To Solve a Word Problem

1. Read the question carefully.

2. If possible, draw a sketch to help visualize the problem.

3. Determine which quantity you are being asked to find. Choose a letter to represent this unknown quantity. Write down exactly what this letter represents. If there is more than one unknown quantity, represent all unknown quantities in terms of this variable.

4. Write the word problem as an equation.

5. Solve the question for the unknown quantity.

6. Answer the question or questions asked.

7. Check the solution in the original stated problem.

Example 1.   Three is added to 5 times a number and the result is 48. Find the number.

Solution.
    1. Let x = unknown number
    2. Write the expression
       "Three is added to 5 times a number"
          $3 + 5x$
       "The result is 48"
          $3 + 5x = 48$
    3. Solve the equation
          $3 + 5x = 48$
         $-3 \qquad\quad -3$

$$\frac{5x}{5} = \frac{45}{5}$$
$$x = 9$$
4. Answer the question
   "The number is 9"
5. Check the solution
   $3 + 5(9) = 48$ ✓

Example 2.   The sum of 2 consecutive odd numbers is 96. Find the numbers.

Solution.
1. Let x = first odd consecutive number. Then $x + 2$ is the next consecutive odd number.
2. Write the equation:
   Sum of 2 consecutive odd numbers is 96
   $x + x + 2 = 96$
3. Solve the equation:
   $$x + x + 2 = 96$$
   $$2x + 2 = 96$$
   $$2x + 2 - 2 = 96 - 2$$
   $$2x = 94$$
   $$\frac{2x}{2} = \frac{94}{2}$$
   $$x = 47$$
4. Answer the question
   $x = 47$, $x + 2 = 49$
5. Check the solution
   $47 + 49 = 96$ ✓

Example 3.   The population of a town increased by 8% one year and reached a new population of 5400. What was the population of the town before the increase?

Solution.
1. Let x = population before increase
2. Write the equation
   Population + 8% of population = new population
   $x + .08 \cdot x = 5400$
3. Solve the equation
   $$1x + .08x = 5400$$
   $$\frac{1.08x}{1.08} = \frac{5400}{1.08}$$
   $$x = 5000$$
4. Check the solution
   $5000 + .08(5000) = 5400$ ✓

Example 4. A collection of nickels, dimes and quarters is worth 3.75. There are twice as many nickels as quarters and 4 times as many dimes as quarters. how many coins of each type are in the collection?

Solution.
1. Let x = number of quarters
   since the number of nickels and dimes are expressed in terms of the number of quarters.
2. Write the equation
   x = number of quarters
   $2x$ = number of nickels
   $4x$ = number of dimes
   (numbers of nickels) · value of a nickel + (number of dimes) · value of a dime + (number of quarters) x value of a quarter = 3.75
   Thus, $(2x)(.05) + 4x)(.10) + x(.25) = 3.75$
3. Solve the equation
   $.1x + .4x + .25x = 3.75$
   $(.1 + .4 + .25)x = 3.75$
   $.75x = 3.75$

   $$x = \frac{3.75}{.75} = 5$$

4. Answer the question
   There are 5 quarters $2(5) = 20$ nickels and $4(5) = 20$ dimes.
5. Check the solution.
   $5(.25) + 10(.05) + 20(.10) = 3.75$

Example 5. A 16 oz container of mouthwash contains 2.4 ounces of hydrogen peroxide. Find the percent by volume of hydrogen peroxide in the mouth wash.

1. Let x = percent of peroxide
2. Write the equation:
   use : total volume x percent of hydrogen peroxide
   = amount of hydrogen peroxide.
   So, $16 \cdot x = 2.4$
   Or $16x = 2.4$
3. Solve the equation:

   $$\frac{16x}{16} = \frac{2.4}{16}$$

   $x = .15$ or 15%
4. Answer the question
   The percent of hydrogen peroxide is 15%.
5. Check the solution
   $16(.15) = 2.4$

Example 6. A taxi cab driver charges $1.10 for the first quarter mile and $.40 for each quarter mile thereafter. If Audra has $10.00 to spend on cab fare, how far can she travel in this cab?

Solution.
1. Let x = number of <u>quarter</u> miles the cab will be driven since the rates are expressed in terms of quarter miles.
2. Write the equation.
   If x = number of quarter miles driven then x - 1 = number of quarter miles driven <u>after</u> the first quarter mile.

   Cab fare = 1.10 + .40(x - 1) or
   10 = 1.10 + .40(x - 1)
3. Solve the equation
   10 = 1.10 + .40(x - 1)
   10 = 1.10 + .40x - .40
   10 = .70 + .40x
   10 - .7 = .40x
   9.3 = .40x

   $\frac{9.3}{.4}$ = x

   23.25 = x
4. Answer the question

   Audra can be driven 23.25 quarter miles or $\frac{23.25}{4}$

   or about 5.8 miles.
5. Check the solution:
   1.10 + (23.25 - 1) .40 = 10.00

Exercise Set 3.3

Set up an algebraic equation that can be used to solve the problem. Solve the equation and answer the question asked.

1. The sum of 2 consecutive even integers is 86. What are the integers?

2. The sum of two numbers is 39. One number is one less than three time the other number. What are the numbers?

3. The sum of three consecutive odd integers is 93. What are the integers?

4. Mr. Fernandez has $4.50 in quarters and dimes. How many quarters and how many dimes does he have?

5. A 32 ounce acid solution has 3.4 ounces of acid. Find the percentage (by volume) of acid in the solution.

6. A small town has a population of 4000. If the population is increasing by 150 each year, how long will it take the town to reach a population of 5800?

7. The cost of renting a car is $25 per day plus $.40 per mile. Find the maximum number of miles the rental car can be driven in one day if $125 is to spent on renting the car.

8. A machine originally costs $10000. If this cost depreciates by the same amount each year for 12 years, the machine will be worth $6640 after this time. Find the amount of depreciation per year for this machine.

9. Allen purchased a pair of slacks at a 20% off sale for a price of $28.79 (before sales tax is added). What was the original price of the slacks?

10. A plumber charges a flat rate of $45 per hour. A second plumber charges a one time fee of $75.00 plus 30.00 per hour. What are the numbers of hours that each plumber can work on a particular job if the cost for completing the job is the same for each plumber? What is the cost?

11. While on vacation in another state, Ms. Walsh purchased some film, postcards, and other miscellaneous items. the bill before tax was $26.00. The bill after tax was $27.82. Fin the local sales tax rate.

12. A soccer coach purchased 9 jerseys at $12.75 and an equal number of pairs of soccer cleats. If the total bill (before sales tax is added) is $258.66, what is the price of a pair of soccer cleats?

Answer for Exercise Set 3.3

1. 42, 44
2. 29, 10
3. 29, 31, 33
4. 12 quarters, 18 dimes
5. 10.63%
6. 12 years
7. 250 miles
8. $280
9. $35.99
10. 5 hrs., cost is $225
11. 7%
12. $15.99

Section 3.4   **Geometric Problems**

Summary:

> **Important Geometric Formulas to recall:**
>
> 1. Perimeter of a rectangle:
> P = 2l + 2w
>
> 2. Area of a rectangle:
> A = l · w
>
> 3. Sum of the interior angles of a triangle is 180°
>
> 4. Sum of the interior angles of a quadrilateral is 360°
>
> 5. Consult the Appendix of your text for other formulas.

Example 1.   The perimeter of a rectangular garden is 38 ft.. The length is 5 feet longer than the width. Find the dimensions of the garden.

Solution.
1. Let x = width. Then length is 5 + x.
2. Write the equation.   P = 2l + 2w
   38 = 2(5 + x) + 2x
3. Solve the equation.   38 = 2(5) + 2x + 2x
   38 = 10 + 4x
   -10   -10

   $$\frac{28}{4} = \frac{4x}{4}$$

   7 = x
4. Answer the question. 7 ft is the width
   7 + 5 = 12 ft is the length
5. Check the solution. 2(7) + 2(12) = 38

Example 2. Find the three angles of a triangle if one angle is 20° more than the smallest angle and the largest angle is 10 more than 4 times the smallest angle.

Solution.

Diagram:

![triangle with angles 4x+10 (top), x+20 (left), x (right)]

1. Let x = smallest angle. Then x + 20 is the middle angle and 4x + 10 is the largest angle.
2. Since the sum of the interior angles of a triangle is 180°, we have x + (x+ 20) + (4x + 10) = 180
3. Solve the equation.

   x + (x + 20) + (4x + 10) = 180

   6x + 30 = 180

   − 30   −30

   $\dfrac{6x}{6} = \dfrac{150}{6}$

4. Answer the question.
   x = 25, x + 20 = 45, and 4x + 10 = 4(25) + 10 = 110°
   The 3 angles are 25°, 45°, 110°
5. Check the solution.
   25° + 45° + 110° = 180°

Example 3. If the 2 larger angles in a parallelogram are 30° less than twice the smaller angles, find the measure of each angle.

Solution.

   FACT: In a parallelogram, opposite angles have the same measure.

1. Let x = measure of the two smaller angles
   Then, 2x − 30 is the measure of the two larger angles.

   FACT: Since a parallelogram is a quadrilateral, the sum of the measures of all 4 angles is 560°.

2. Write the equation. x + x + (2x − 30) + (2x − 30)= 300
3. Solve the equation. 2x + 4x − 60   = 360
                        6x − 60   = 360
                           + 60   + 60
                        6x   = 420
                        x   = 70

4. Answer the question. The measure of the 2 smaller angles is 70° while the 2 larger angles each measure 2(70) − 30 = 110°.

Example 4: A bookshelf (shown) is to have 3 shelves, including the top. The height of the bookshelf is to be 1 foot less than twice the width. Find the dimensions of the bookshelf if 33 feet are available.

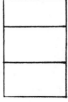

Solution.

1. Let x = width of bookshelf.
   Then, 2x - 1 is the height of the bookshelf.

2. Write the equation.
   (3 shelves) + (2 sides) = total lumber available
   $$3x + 2(2x - 1) = 33$$
3. Solve the equation.
   $$3x + 4x - 2 = 33$$
   $$7x - 2 = 33$$
   $$7x = 35$$
   $$x = 5$$

4. Answer the question.
   The width of the shelf is 5 feet. The height of the shelf is 2(5) - 1 = 9 ft.
5. Check the solution.
   $$3(5) + 2(9) = 15 + 18 = 33$$

Exercise Set 3.4

Write an algebraic equation for each of the following problems. Solve the equation and answer the question.

1. The perimeter of a room is 52 feet. The length of the room is four feet less than twice the width. Find the room's dimensions.

2. One angle of a right triangle is twice as large as the smallest angle. Find the measure of the three angles of this triangle.

3. The perimeter of an isosceles triangle is 23 meters. Find the length of all 3 sides if the 2 sides having the same length are 1 meter less than twice the length of the base.

4. Find the cost of installing baseboard along the base of the walls of a rectangular room which measures 12 ft x 18 ft if the cost of the baseboard is 30¢ per linear foot.

5. In a triangle, one angle is twice as large as another angle and the third angle is $40°$ larger than the smallest angle. Find the measure of all 3 angles.

6. In a parallelogram, the measure of the 2 largest angles is 20° more than three times the measure of the 2 smaller angles. Find the measure of all the angles.

7. A bookshelf is to consists of 5 shelves (including the top) and 2 sides which measure 1.5 times the length of each shelf. If 64 board feet of lumber are available, what are the dimensions of the bookshelf?

8. Moses wants a garden whose length is 4 meters longer than its width. The perimeter of the garden is to be 70 ft. What are the dimensions of the garden?

9. If angles A and B are supplementary, and angle B is 10° more than angle A, find the measure of each angle.

10. Angles A and B are complementary. Angle B is 9° more than twice the measuring of angle A. Find the measure of each angle.

11. If triangle ABC is equilateral and its perimeter is 25.5 ft, find the length of each side.

12. The volume, V, of a box is given by $V = l \cdot w \cdot h$. What happens to the volume if the length, width and height are all tripled?

13. In the equation, $A = l \cdot w$, what happens to the area if the length is quadrupled and the width is halved?

14. In the formula $A = s^2$, what happens to the area if the length of a side, s, is tripled?

Answer to Exercise Set 3.4

1. 10' x 16'
2. 30°, 60°, 90°
3. 5 ft, 9 ft, 9 ft
4. $18.00
5. 35°, 70°, 75°
6. 40°, 40°, 140°, 140°
7. 8 ft x 12 ft
8. w = 15.5 ft, l = 19.5 ft
9. ∠A = 85°, ∠B = 63°
10. ∠A = 27°, ∠B = 63°
11. 8.5 ft
12. The volume becomes 27 times as great.
13. The area becomes twice as great.
14. The area becomes 9 times as great.

Practice Test Chapter 3

1. Given $m = \dfrac{a+b}{2}$, find b when m is 36 and a is 16.

2. Find H if $V = \dfrac{1}{3} BH$, B = 240 and V = 960.

3. Solve for y then find the value of y for the given value of x.
   3x - 4y = 12; x = 8

4. Solve P = 2l + 2w for l.

5. Solve $F = \dfrac{5}{9} C + 32$ for C.

6. Solve $V = \dfrac{4}{3} \pi r^2 h$.

7. Solve 2x - 5y = 10 for y.

Write an algebraic equation for each problem. Solve the equation and then answer the question.

8. The sum of 2 consecutive odd integers is 164. Find the integers.

9. What is the cost of a car before tax if the total cost of the car including a 7% sales tax is $8744?

10. The first side of a triangle is 5 meters less than the second side, and the third side is twice as long as the first side. If the perimeter is 33 meters, find the lengths of all 3 sides of the triangle.

11. The cost of renting a truck is $50.00 plus $22.00 per day. If Rodriguez has $120.00 to spend on truck rental, for how many days can he rent the truck?

12. The 2 larger angles of a parallelogram are 30° more than twice the measure of the smaller angles. Find the measure of each angle of the parallelogram.

13. A photocopier is on sale for $460.00. Copies can be made at a local copy center for $.04 a copy. How many copies can be made for the copying cost to equal the cost of the copier?

14. You have $20.00 to spend at the grocery store. Your purchases

so far total $13.10   How many large bags of corn chips can you buy at $2.30 per bag with your remaining money?

Answers to Practice Test Chapter 3

1. b = 56
2. H = 12

3. y = 3
4. $l = \dfrac{p - 2w}{2}$ or $l = \dfrac{1}{2}(p - 2w)$

5. $C = \dfrac{9}{5}(F - 32)$
6. $h = \dfrac{3V}{4\pi r^2}$

7. $y = \dfrac{2}{5}x - 2$
8. 81, 83

9. $8,200
10. 7m, 12m, 14m

11. 3 days, he has $4.00 to spare
12. 50°, 50°, 130°, 130°
13. 11,500

14. 3

Set 4.1   **Exponents**

Summary:

> In the expression x^n, x is referred to as the **base** and n is called the **exponent**. x^n is read "x to the nth power."
>
> $$x^n = \underbrace{x \cdot x \ldots x}_{n \text{ factors of } x}$$

Example 1. Write $x \cdot x \cdot x \cdot y \cdot y$ using exponents

$$\underbrace{x \cdot x \cdot x}_{3 \text{ factors}} \cdot \underbrace{y \cdot y}_{2 \text{ factors}} = x^3 y^2$$

Summary:

> **Product Rule of Exponents**
>
> $$x^m \cdot x^n = x^{m+n}$$

Example 2. Multiply using the product rule

a) $y^3 \cdot y^2$ $y^3 \cdot y^2 = y^{3+2} = y^5$

Note: We could write $y^3 \cdot y^2$ or $\underbrace{y \cdot y \cdot y}_{3 \text{ factors}} \cdot \underbrace{y \cdot y}_{2 \text{ factors}} = y^5$

b) $4^2 \cdot 4^7$ $4^2 \cdot 4^7 = 4^{2+7} = 4^9$

c) $a^5 \cdot a$ $a^5 \cdot a = a^5 \cdot a^1 = a^{5+1} = a^6$

d) $a^2 \cdot b^4$ $a^2 b^4$ bases are not the same so product rule cannot be applied

Summary:

Quotient Rule

$$\frac{x^m}{x^n} = x^{m-n} \qquad x \neq 0$$

Example 3. Divide using the quotient rule

a) $\dfrac{x^6}{x^4}$ $\qquad\qquad \dfrac{x^6}{x^4} = x^{6-4} = x^2$

Note: We could write $\dfrac{x^6}{x^4}$ or $\dfrac{x \cdot x \cdot x \cdot x \cdot x \cdot x}{x \cdot x \cdot x \cdot x}$

$\qquad\qquad\qquad\qquad\qquad\qquad\qquad = x \cdot x = x^2$

b) $\dfrac{y^8}{y^2}$ $\qquad\qquad \dfrac{y^8}{y^2} = y^{8-2} = y^6$

c) $\dfrac{7^5}{7^4}$ $\qquad\qquad \dfrac{7^5}{7^4} = 7^{5-4} = 7^1 = 7$

d) $\dfrac{b^5}{b}$ $\qquad\qquad \dfrac{b^5}{b} = \dfrac{b^5}{b^1} = b^{5-1} = b^4$

e) $\dfrac{x^3}{y^2}$ $\qquad\qquad \dfrac{x^3}{y^2}$ bases are not the same so quotient rule cannot be used

Note: $\dfrac{5^4}{5^2} \neq \left(\dfrac{5}{5}\right)^{4-2}$. Do not divide out the bases

Summary:

> In this section, to simplify an expression when the numerator and denominator have the same base and the greater exponent is in the denominator, we divide out common factors.

Example 4. Simplify by dividing out a common factor is both the numerator and denominator.

a) $\dfrac{a^2}{a^5}$ $\qquad$ $\dfrac{a^2}{a^5} = \dfrac{a^2}{a^2 \cdot a^3} = \dfrac{1}{a^3}$

b) $\dfrac{7^5}{7^6}$ $\qquad$ $\dfrac{7^5}{7^6} = \dfrac{7^5}{7^5 \cdot 7^1} = \dfrac{1}{7^1} = \dfrac{1}{7}$

Summary

> **Zero Exponent Rule**
>
> $x^0 = 1, \; x \neq 0$

Example 5: Simplify each expression

a) 4^0 $\qquad\qquad$ $4^0 = 1$

b) a^0 $\qquad\qquad$ $a^0 = 1$

c) $2a^0$ $\qquad\qquad$ $2(a^0) = 2(1) = 2$

d) $(5a)^0$ $\qquad\qquad$ $(5a)^0 = 1$

Summary:

> **Power Rule for Exponents**
>
> $(x^m)^n = x^{mn}$

Example 5. Simplify each term

a) $(y^2)^4$ $(y^2)^4 = y^{2 \cdot 4} = y^8$

b) $(3^3)^5$ $(3^3)^5 = 3^{3 \cdot 5} = 3^{15}$

Summary:

> **Expanded Power Rule for Exponents**
>
> $\left(\dfrac{ax}{by}\right)^m = \dfrac{a^m x^m}{b^m y^m}$ $b \neq 0, \ y \neq 0$

Example 6. Simplify each expression

a) $(3y)^3$ $(3y)^3 = 3^3 y^3 = 27y^3$

b) $(-a)^4$ $(-a^4) = (-1a)^4 = (-1)^4(a^4) = 1 \cdot a^4 = a^4$

c) $(4cd)^2$ $(4cd)^2 = 4^2 c^2 d^2 = 16c^2 d^2$

d) $\left(\dfrac{-4y}{3x}\right)^3$ $\left(\dfrac{-4y}{3x}\right)^3 = \dfrac{(-4)^3 y^3}{3^3 \cdot x^3} = \dfrac{-64y^3}{27x^3}$

Summary:

> Whenever we have an expression raised to a power, it helps to simplify the expression in parentheses before using the expanded power rule.

Example 7. Simplify each expression

a) $\left(\dfrac{9x^6y^3}{3x^2y^2}\right)^2$

First, simplify expression within parentheses

$$\left(\dfrac{9x^6y^3}{3x^2y^2}\right)^2 = (3x^4y)^2$$

Now, apply expanded power rule

$$(3x^4y)^2 = 3^2x^8y^2 = 9x^8y^2$$

b) $(4x^2y^5)^3(x^3y^2)^2$

Here, cannot simplify expression in parentheses, so use expanded power rule format.

$$(4x^2y^5)^3 = 4^3 \cdot x^6y^{15} = 64x^6y^{15}$$

Now apply the product rule

$$64x^6y^{15} \cdot x^6y^4 = 64x^{6+6} \cdot y^{15+4} = 64x^{12}y^{19}$$

Summary of the Rules of Exponents Presented in this Chapter

1. $x^m \cdot x^n = x^{m+n}$ Product Rule

2. $\dfrac{x^m}{x^n} = x^{m-n}$ Quotient Rule

3. $x^0 = 1, \; x \neq 0$ Zero Exponent Rule

4. $(x^m)^n = x^{mn}$ Power Rule

5. $\left(\dfrac{ax}{by}\right)^m = \dfrac{a^m x^m}{b^m y^m}, \; b \neq 0, \; y \neq 0$ Expanded Power Rule

Exercises Set 4.1

Simplify the following:

1. $c^4 \cdot c^2$
2. $5^3 \cdot 5^5$
3. $(-x)^2(-x)^3$
4. $r^8 \cdot r^5$
5. $\dfrac{u^6}{u^2}$
6. $\dfrac{n^3}{n^2}$
7. $\dfrac{8^7}{8}$
8. $\dfrac{z^{12}}{z^3}$
9. $(3^5)^2$
10. $(y^3)^7$
11. $(4d^3)^3$
12. $(-3k)^3$
13. $(-1.5)^0$
14. $(6x^5)^0$
15. $\dfrac{x^3 y^8}{x^5 y^5}$
16. $\left(\dfrac{2x^3}{y^2}\right)^2$
17. $\left(\dfrac{a^2 b^3}{a^5 b}\right)^4$
18. $(xy)^2(2x^2 y^2)^4$
19. $(2a^3 b)^4 (a^2 b^3)$
20. $\left(\dfrac{20x^3 y^7}{5xy^9}\right)^2$

Answers to Exercises Set 4.1

1. c^6
2. 5^8
3. $-x^5$
4. r^{13}
5. u^4
6. n
7. 8^6
8. z^9
9. 3^{10}
10. y^{21}
11. $64d^9$
12. $-27k^3$

13. 1   14. 1   15. $\dfrac{y^3}{x^2}$

16. $\dfrac{4x^6}{y^4}$ 17. $\dfrac{n^8}{a^{12}}$ 18. $16x^{10}y^{10}$

19. $16a^{14}b^7$ 20. $\dfrac{16x^4}{y^4}$

Set 4.2 Negative Exponents

Summary:

> **Negative Exponent Rule**
>
> $x^{-m} = \dfrac{1}{x^m}$, $x \neq 0$

Example 1. Use the negative exponent rule to write each expression with positive exponents.

a) $\dfrac{1}{y^{-5}}$ $\qquad\qquad y^{-5} = \dfrac{1}{y^5}$

b) 3^{-2} $\qquad\qquad 3^{-2} = \dfrac{1}{3^2}$

Note: Raising a number to a negative exponent does not automatically makes the value of the expression negative.

c) $\dfrac{1}{a^{-5}}$ $\qquad\qquad \dfrac{1}{a^{-5}} = \dfrac{1}{\frac{1}{a^5}} = \dfrac{1}{1} \cdot \dfrac{a^5}{a^5} = a^5$

d) $\dfrac{1}{6^{-1}}$ $\qquad\qquad \dfrac{1}{6^{-1}} = \dfrac{1}{\frac{1}{6}} = \dfrac{1}{1} \cdot \dfrac{6^1}{1} = 6$

Here are more examples that involve some of preceeding rules of exponents along with the negative exponent rule.

Example 2. Simplify

a) $(a^3)^{-4}$ $\qquad\qquad (a^3)^{-4} = a^{-12}$ by the product rule

$\qquad\qquad\qquad\qquad = \dfrac{1}{a^{12}}$ by the negative exponent rule

b) $(7^{-3})^2$ $\qquad\qquad (7^{-3})^2 = 7^{-6}$ by the product rule

$\qquad\qquad\qquad\qquad = \dfrac{1}{7^6}$ by the negative exponent rule

Example 3. Simplify. Write without negative exponents

a) $b^{-5} \cdot b^2$

$b^{-5} \cdot b^2 = b^{-5+2}$ by the product rule
$= b^{-3}$

$= \dfrac{1}{b^3}$

by the negative exponent rule

c) $z^{-4} \cdot z^{-3}$

$z^{-4} \cdot z^{-3} = z^{-4+(-3)}$
$= z^{-7}$

$= \dfrac{1}{z^7}$

by the negative exponent rule

d) $\dfrac{n^{-6}}{n^{-2}}$

$\dfrac{n^{-6}}{n^{-2}} = n^{-6-(-2)}$ by the quotient rule

$= n^{-4}$

$= \dfrac{1}{n^4}$ by the negative exponent rule

e) $\dfrac{12x^2 y^{-3}}{4x^{-1} y^3}$

$\dfrac{12x^2 y^{-3}}{4x^{-1} y^3} = \dfrac{12}{4} \cdot \dfrac{x^2}{x^{-1}} \cdot \dfrac{y^{-3}}{y^3}$

$= 3x^3 y^{-6}$

$= \dfrac{3x^3}{y^6}$

f) $(2x^{-3})^{-4}$

$(2x^{-3})^{-4} = 2^{-4} x^{12}$ by expanded power rule

$= \dfrac{x^{12}}{2^4}$ by negative exponent rule

$= \dfrac{x^{12}}{16}$

Summary of Rules of Exponents

1. $x^m \cdot x^n = x^{m+n}$
   product rule

2. $\dfrac{x^m}{x^n} = x^{m-n}$, $x \neq 0$
   quotient rule

3. $x^0 = 1$, $x \neq 0$
   zero exponent rule

4. $(x^m)^n = x^{mn}$
   power rule

5. $\left(\dfrac{ax}{by}\right)^m = \dfrac{a^m x^m}{b^m y^m}$, $b \neq 0$, $y \neq 0$
   expanded power rule

6. $x^{-m} = \dfrac{1}{x^m}$, $x \neq 0$
   negative exponent rule

Exercise Set 4.2

Simplify each of the following. Write results without negative exponents.

1. a^{-3}
2. $\dfrac{1}{4^{-3}}$
3. $(3x)^{-4}$
4. $(5a)^{-1}$
5. $y^{-2} \cdot y^{-4}$
6. $r^3 \cdot r^{-7}$
7. $\dfrac{k^6}{k^8}$
8. $\dfrac{u^{-7}}{u^{-5}}$
9. $(2x^{-1}y^3)(5xy^{-5})$
10. $\dfrac{18a^{-3}b^5}{6a^2b^{-8}}$
11. $(4c^{-3}d^2)^{-2}$
12. $(2x^5y^{-2})^{-4}$

Answers for exercise Set 4.2

1. $\dfrac{1}{a^3}$
2. 4^3
3. $\dfrac{1}{81x^4}$ or $\dfrac{1}{3^4 x^4}$
4. $\dfrac{1}{5a}$
5. $\dfrac{1}{y^6}$
6. r^4
7. $\dfrac{1}{k^2}$
8. $\dfrac{1}{u^2}$
9. $\dfrac{10}{y^2}$
10. $\dfrac{3b^{13}}{a^5}$
11. $\dfrac{c^6}{16d^4}$
12. $\dfrac{y^8}{16x^{20}}$

Section 4.3 **Scientific notation**

> **To Write a Number in Scientific Notation**
>
> 1. Move the decimal point to the right of the first nonzero digit. This will give a number greater than or equal to 1 and less than 10.
>
> 2. Count the number of places you moved the decimal to obtain the number in step 1. If the original number was 10 or greater, the count is considered positive. If the original number was less than 1, the count is to be considered negative.
>
> 3. Multiply the number obtained in step 1 by 10 raised to the count (power 0 found in step 2.

Example 1: Write the following number in scientific notation

a) 200,200

200,200 means 200,200
$200,200 = 2.002 \times 10^5$

Note: We moved decimal point 5 places and 200,200 is greater than 10, so the power of 10 is positive 5.

b) 0.0056

0.0056 means 0.0056
$0.0056 = 5.6 \times 10^{-3}$

Note: We moved the decimal point 3 places and 0.0056 is less than 10, so the power of 10 is negative 3.

c) 40,000,000

40,000,000 means 40,000,000
$40,000,000 = 4 \times 10^7$

Summary:

> **To Convert a Number from Scientific Notation to Decimal Form**
>
> 1. Obtain the exponent of the power of 10
>
> 2a. If the exponent is positive, move the decimal in the number (greater than or equal to 1 and less than 10) to the right the same number of places as exponent. It may be necessary to add zeros to the number. This will result in a number greater than or equal to 10.
>
> 2b. If the exponent is 0, do not move the decimal point. Drop the factor 10^0 since it equals 1. This will result in a number greater than or equals to 1.
>
> 2c. If the exponent is negative, move the decimal point in the number to the left the same number of places as the exponent (dropping the negative sign). It may be necessary to add zeros. This will result in a number less than 1.

Example 2. Write each number without exponents

a) 5.53×10^3

Move decimal 3 places to right gives
$$5.53 \times 10^3 = 5.53 \times 1000 = 5530$$

b) 6.25×10^{-6}
Move decimal 6 places to left gives
$$6.25 \times 10^{-6} = 0.00000625$$

c) 3.5×10^0
Since exponent is 0, drop the factor 10^0. So, $3.5 \times 10^0 = 3.5$

We can use the rule of exponents presented in Section 4.1 and 4.2 when working with numbers written scientific notation.

Example 3. Multiply $(2.5 \times 10^8)(1.2 \times 10^{-4})$

$$
\begin{aligned}
(2.5 \times 10^8)(1.2 \times 10^{-4}) &= (2.5 \times 1.2)(10^8 \times 10^{-4}) \\
&= 3 \times 10^{8-4} \\
&= 3 \times 10^4 \text{ or } 30,000
\end{aligned}
$$

Example 4. Divide $\dfrac{(6 \times 10^{-8})}{(2.5 \times 10^{-11})}$

$$
\begin{aligned}
\dfrac{(6 \times 10^{-8})}{(2.5 \times 10^{-11})} &= \dfrac{(6)}{(2.5)} \times \dfrac{(10^{-8})}{(10^{-11})} \\
&= 2.4 \times 10^{-8-(-11)} \\
&= 2.4 \times 10^3 \text{ or } 2,400
\end{aligned}
$$

Example 5. Perform the indicated operation by first converting each number to scientific notation. Write the answer in scientific notation form.

$$
\begin{aligned}
\dfrac{(2,000,000)(13,000)}{0.00065} &= \dfrac{(2 \times 10^6)(1.3 \times 10^4)}{(6.5 \times 10^{-4})} \\
&= \dfrac{2 \times 1.3 \times 10^6 \times 10^4}{6.5 \times 10^{-4}} \\
&= \dfrac{2.6 \times 10^6 \times 10^4}{6.5 \times 10^{-4}} \\
&= \dfrac{2.6 \times 10^{10}}{6.5 \times 10^{-4}} \\
&= \left(\dfrac{2.6}{6.5}\right) \times \left(\dfrac{10^{10}}{10^{-4}}\right) \\
&= 0.4 \times 10^{10-(-4)} \\
&= 0.4 \times 10^{14} \\
&= 4 \times 10^{-1} \times 10^{14} \\
&= 4 \times 10^{13}
\end{aligned}
$$

Example 6. Use scientific notation to determine the answer to the following. Write the answer without exponents.

A light year is approximately 5.86 trillion miles (5,860,000,000,000 miles). The diameter of the Milky Way Galaxy is approximately 150,000 light years. What is the diameter of the Milky Way Galaxy in miles?

$$
\begin{aligned}
(5{,}860{,}000{,}000{,}000)(150{,}000) &= (5.86 \times 10^{12})(1.5 \times 10^{5}) \\
&= 5.86 \times 1.5 \times 10^{12} \times 10^{5} \\
&= 8.79 \times 10^{12} \times 10^{5} \\
&= 8.79 \times 10^{12 + 5} \\
&= 8.79 \times 10^{17} \\
&= 879{,}000{,}000{,}000{,}000{,}000
\end{aligned}
$$

This number is read "879 quadrillion."

Exercises Set 4.3

Express each number in scientific notation:

1. 800,000
2. 4,000,000
3. 0.0000075
4. 60
5. 0.000213
6. 3

Express each number without exponents:

7. 4.25×10^{7}
8. 3.1×10^{-5}
9. 8×10^{4}
10. 3.14×10^{6}
11. 4.653×10^{0}
12. 2.718×10^{-2}

Perform the indicated operation and express each number without exponents:

13. $(2.5 \times 10^{5})(1.4 \times 10^{4})$
14. $\dfrac{(8 \times 10^{7})}{(4 \times 10^{3})}$
15. $(5 \times 10^{6})(8 \times 10^{-10})$
16. $\dfrac{(3 \times 10^{-2})}{(4 \times 10^{-8})}$
17. $(2.5 \times 10^{8})(4 \times 10^{-9})$

Perform the indicated operation by first converting each number to scientific notation. Write the answer in scientific notation:

18. $\dfrac{(0.0072)(0.000003)}{40,000}$

19. $\dfrac{(21,000)(400,000)}{(150)}$

20. A rocket travels at the rate of 25,000 miles per hour. The distance to Neptune is 2,000,000,000 miles. How many hours will it take the rocket to reach Neptune?

Answers to Exercise Set 4.3

1. 8×10^5
2. 4×10^6
3. 7.5×10^{-6}
4. 6×10^1
5. 2.13×10^{-5}
6. 3×10^0
7. 42,500,000
8. 0.000031
9. 80,000
10. 3,140,000
11. 4.653
12. 0.02718
13. 3,500,000,000
14. 20,000
15. 0.0004
16. 750,000
17. 1
18. 5.4×10^{-13}
19. 5.6×10^7
20. 80,000 hours

Section 4.4   Addition and Subtraction of Polynomials

Summary:

> A **polynomial in x** is an expression containing the sum of a finite number of terms of the form ax^n, for any real number **a** and any whole number **n**.

Examples of Polynomials

$4x$

$\frac{2}{3}x^2 - 5$

$x^3 - \frac{1}{2}x + 4$

Not polynomials

$3x^{1/3}$

(fractional exponent)

$x^3 + x^{-2}$

(negative exponent)

$\frac{4}{x} + x^3 \qquad \frac{4}{x} = 4x^{-1}$

(negative exponents)

Summary:

> A polynomial is written in **descending order** or **descending power of the variable** when the exponent on the variable decreases from left to right.

An example of a polynomial in descending order is

$2x^3 + 6x^2 - 8x + 5$

Summary:

> A **monomial** is a one-term polynomial.
>
> A **binomial** is a two-termed polynomial.
>
> A **trinomial** is a three-termed polynomial.

Summary:

> The **degree of a term** of a polynomial in one variable is the exponent on the variable of that term.

Examples:

Terms	Degree of Terms
$6x^3$	third
$3a^7$	seventh
$-2x$	first ($-2x = -2x^1$)
5	zero ($5 = 5x^0$)

Summary:

> The **degree of a polynomial** in one variable is the same as that of its highest-degree term.

Examples:

Polynomial	Degree of the Polynomial
$4x^5 - 6x - 3$	fifth ($4x^5$ is highest-degree term)
$8x + 1$	first ($8x$ is highest-degree term)
15	zero ($15 = 15^0$)

Recall, like terms differ only in their numerical coefficients.

Examples of Like Terms

6	-3
$7x$	$3x$
$-3y^2$	$6y^2$
$2ab^2$	$-5ab$

Summary:

> **To add polynomials,** combine the like terms of the polynomial.

Example 1. Simplify the following polynomials:

a) $(-2x^2 + 5x - 5) + (6x^2 - 8x + 3)$

$(-2x^2 + 5x - 5) + (6x^2 - 8x + 3)$
$-2x^2 + 5x - 5 + 6x^2 - 8x + 3$ Remove parentheses
$-2x^2 + 6x^2 + 5x - 8x - 5 - 3$ Rearange terms
$4x^2 - 3x - 8$ Combine like terms

b) $(2x^3 + 6x - 2) + (x^2 - 3x + 7)$

$(2x^3 + 6x - 2) + (x^2 - 3x + 7)$ Remove parentheses
$2x^3 + x^2 + 6x - 3x - 2 + 7$ Rearange terms
$2x^3 + x^2 + 3x + 5$ Combine like terms

Summary

> **To Add Polynomials in Columns**
>
> 1. Arrange polynomials in descending order one under the other with like terms in the same column.
>
> 2. Add the terms in each column.

Example 2. Add the following polynomials using columns.

a) $-2x^2 + 7x - 6$ and $3x^2 - 3x + 9$

$-2x^2 + 7x - 6$
$\underline{3x^2 - 3x + 9}$
$x^2 + 4x + 3$

b) $3y^3 + 6y^2 + 3$ and $3y^2 - 7y - 6$

$$\begin{array}{r} 3y^3 + 6y^2 + 0y + 3 \\ \underline{3y^2 - 7y - 6} \\ 3y^3 + 9y^2 - 7y - 3 \end{array}$$ (Note 0y for the missing term)

Summary

To Subtract Polynomials

1. Remove parentheses (This will have the effect of changing the sign of every term within the parentheses of the polynomial being subtracted.)

2. Combine like terms.

Example 3. Simplify the following polynomials

a) $(6a^2 + 3a - 10) - (2a^2 - 5a - 7)$

$(6a^2 + 3a - 10) - (2a^2 - 5a - 7)$
$6a^2 + 3a - 10 - 2a^2 + 5a + 7$ Remove parentheses (Change the sign of each term being subtracted)
$6a^2 - 2a^2 + 3a + 5a - 10 + 7$ Rearrange terms
$4a^2 + 8a - 3$ Combine like terms

Note: A common student error is to fail to change the sign of <u>every</u> term in the polynomial being subtracted.

b) $(-x^3 + 3x - 6) - (x^2 - 6x + 2)$

$(-x^3 + 3x - 6) - (x^2 - 6x + 2)$
$-x^3 + 3x - 6 - x^2 + 6x - 2$ Remove parentheses
$-x^3 - x^2 + 3x + 6x - 6 - 2$ Rearrange terms
$-x^3 - x^2 + 9x - 8$ Combine like terms

Summary:

> **To Subtract Polynomials in Columns**
>
> 1. Write the polynomial being subtracted below the polynomial from which it is being subtracted. List like terms in the same column.
>
> 2. **Change the sign of each term** in the polynomial being subtracted.
> (This step can be done mentally if you like.)
>
> 3. Add the terms in each column.

Example 4. Subtract using columns.

a) $(3x^2 - 7x + 11) - (10x^2 + 8x - 5)$

$$\begin{array}{l} 3x^2 - 7x + 11 \\ \underline{10x^2 + 8x - 5} \end{array} \quad \text{Align like terms}$$

$$\begin{array}{l} 3x^2 - 7x + 11 \\ \underline{-10x^2 - 8x + 5} \\ -7x^2 - 15x + 16 \end{array} \quad \begin{array}{l} \text{Change all signs} \\ \text{Add} \end{array}$$

b) $(5x^3 + 6x - 1) - (3x^2 - 8x - 2)$

$$\begin{array}{l} 5x^3 + 0x^2 + 6x - 1 \\ \phantom{5x^3 + {}} 3x^2 - 8x - 2 \end{array} \quad \begin{array}{l} \text{Note } 0x^2 \text{ for missing term} \\ \text{Align like terms} \end{array}$$

$$\begin{array}{l} 5x^3 + 0x^2 + 6x - 1 \\ \underline{\phantom{5x^3 + {}} - 3x^2 + 8x + 2} \\ 5x^3 - 3x^2 + 14x + 1 \end{array} \quad \begin{array}{l} \text{Change all signs} \\ \text{Add} \end{array}$$

Exercise Set 4.4

Indicate the expressions that are polynomials. If the polynomial have a specific name - for example monomial or binomial - give that name.

1. $4x^3 + 6x$
2. $4x^{1/3} - 6x$
3. $\frac{7}{x} - 1$
4. 15

Express each polynomial in descending order. If the polynomial is already in descending order, so states. Give the degree of each polynomial.

5. $2b + b^2 - 8$
6. $5x$
7. $-2x^3 + 1 - x^4$
8. -3

Add as indicated.

9. $(3u^2 - 7u + 3) + (2u^2 + 2u - 6)$
10. $(6a^2 + 3a - 2) + (4a^2 - 3a - 6)$
11. $(6x^2 - 7x + 1) + (2x^2 - 8)$
12. $(4x^4 - 1) + (4x^4 + 2)$
13. $(-b + 2b^2 - 1) + (b^2 - 6b + 2)$
14. Add in columns. $(3x^2y - 5xy + 5y^2)$ and $(-5x^2y + 7xy - 8y^2)$

Subtract as indicated.

15. $(5x^2 - 3x - 7) - (8x^2 + 2x - 4)$
16. $(-5a^2 + 10a) - (-8a^2 - 11a)$
17. $(2k^2 - 6k + 3) - (5k^2 + 6k + 2)$
18. $(k^3 - k) - (k^2 - 2k - 1)$
19. Subtract in columns. $(10y^3 - 7y^2 - 2y) - (8y^3 + 6y^2 - 3y)$
20. Subtract in columns. $(6y^3 + 10y - 1) - (8y^2 + 3 - 11y)$

Answers to Exercises Set 4.4

1. Binomial
2. not a polynomial
3. not a polynomial
4. Monomial
5. $b^2 + 2b - 8$, second degree
6. $5x$, first degree
7. $-x^4 - 2x^3 + 1$, fourth degree
8. -3, zero degree
9. $5u^2 - 5u - 3$
10. $10a^2 - 8$
11. $8x^2 - 7x + 1$
12. $8x^4 + 1$
13. $3b^2 - 7b + 1$
14. $-2x^2y + 2xy - 3y^2$
15. $-3x^2 - 5x - 3$
16. $3a^2 + 21a$
17. $-3k^2 + 1$
18. $k^3 + k + 1$
19. $2y^3 - 13y^2 + y$
20. $6y^3 - 8y^2 + 21y - 1$

Set 4.5   Multiplication of Polynomials

Summary:

> To multiply a monomial by a monomial, multiply their coefficients and use the product rule of exponents to determine the exponent or variable.

Example 1.   Multiply the following.

a) $(5x^4)(3x^3) = (5 \cdot 3)(x^4 \cdot x^3)$
$= 15 \cdot x^{4+3}$
$= 15x^7$

b) $(-2x^2y^3)(-3x^4y) = (-2)(-3)(x^2 \cdot x^4 \cdot y^3 \cdot y^1)$
$= 6 \cdot x^{2+4} \cdot y^{3+1}$
$= 6 \cdot x^6y^4$

Summary:

> To multiply a polynomial by a monomial, we use the distribution property.
>
> $a(b = c) = ab + ac$
> $a(b + c + d + \ldots n) = ab + ac + ad + \ldots an$

Example 2.   Multiply the following.

a) $3a(2a^2 - 5) = (3a)(2a) + (3a)(-5)$
$= 6a^2 - 15$

b) $-2c(c^2 - 4c + 2) = (-2c)(c^2) + (-2c)(-4c) + (-2c)(2)$
$= -2c^3 + 8c - 4c$

c) $5y^3(y^2 - 3y - 1) = (5y^3)(y^2) + (5y^3)(-3y) + (5y^3(-1)$
$= 5y^5 - 15y^4 - 5y^3$

Summary:

> To multiply a binomial by a binomial use the distributive property in the following manner:
>
> $$(a + b)(c + d) = (a + b)c + (a + b)d$$
> $$= ac + bc + ad + bd$$

Example 3. Multiply $(2x - 3)(x + 4)$

$$\begin{aligned}(2x - 3)(x + 4) &= (2x - 3)(x) + (2x - 3)(4) \\ &= (2x)(x) + (-3)(x) + (2x)(4) + (-3)(4) \\ &= 2x^2 - 3x + 8x - 12 \\ &= 2x^2 + 5x - 12\end{aligned}$$

Summary:

> ### F O I L
>
> Consider $(a + b)(c + d)$
>
> F stands for first
>
> $(a + b)(c + d)$ product ac
>
> O stands for outer
>
> $(a + b)(c + d)$ product ad
>
> I stands for inner
>
> $(a + b)(c + d)$ product bc
>
> L stands for last
>
> $(a + b)(c + d)$ product bd
>
> The product of the two binomials is the sum of these four products.
>
> $(a+b)(c+d) = ac + ad + bc + bd$

Example 4. Use the FOIL method to multiply:

a) $(3x + 1)(2x - 3)$

$$(3x + 1)(2x - 3)$$

$$\overset{F}{(3x)(2x)} + \overset{O}{(3x)(-3)} + \overset{I}{(1)(2x)} + \overset{L}{(1)(-3)}$$
$$6x^2 - 9x + 2x - 3$$
$$6x^2 - 7x - 3$$

b) $(2x - 3)(2x + 3)$

$$(2x)(2x) + (2x)(3) + (-3)(2x) + (-3)(3)$$
$$4x^2 + 6x - 6x - 9$$
$$4x^2 - 9$$

Summary:

Product of the Sum and Differences of the Same two Terms

$(a + b)(a - b) = a^2 - b^2$

Example 5. Use the rules for finding the product of the sum and difference of two quantities to multiply each expression.

a) $(x - 5)(x + 5)$ where $a = x$ and $b = 5$

$$(a + b)(a - b) = a^2 - b^2$$
$$(x - 5)(x + 5) = (x)^2 - (5)^2$$
$$= x^2 - 25$$

b) $(3x - 2y)(3x + 2y)$ where $a = 3x$ and $b = 2y$

$$(a + b)(a - b) = a^2 - b^2$$
$$(3x\ 2y)(3x + 2y) = (3x)^2 - (2y)^2$$
$$= 9x^2 - 4y^2$$

Summary:

> **Square of Binomial Formulas**
>
> $(a + b)^2 = (a + b)(a + b) = a^2 + 2ab + b^2$
>
> $(a - b)^2 = (a - b)(a - b) = a^2 - 2ab + b^2$

Example 6. use the square of a binomial formula to multiply each expression.

a) $(x + 6)^2$ Where $a = x$, $b = 6$

$$(a + b)^2 = a^2 + 2 \cdot a \cdot b + b^2$$
$$(x + b)^2 = (x)^2 + 2(x)(6) + (6)^2$$
$$= x^2 + 12x + 36$$

b) $(2t - 5c)^2$ Where $a = 2t$, $b = 5c$

$$(a - b)^2 = a^2 - 2 \cdot a \cdot b + b^2$$
$$(2t - 5c)^2 = (2t)^2 - 2(2t)(5c) + (5c)^2$$
$$= 4t^2 - 20tc + 25c^2$$

c) $(k - 4)(k - 4)$

$(k - 4)(k - 4) = (k - 4)^2$ Where $a = k$, $b = 4$

$$(a - b)^2 = a^2 - 2 \cdot a \cdot b + b^2$$
$$(k - 4)^2 = k^2 - 2(k)(4) + 4^2$$
$$= k^2 - 8k + 16$$

Note: $(a + b)^2 \ne a^2 + b^2$. A common student error is to forget the middle term $2ab$.

Summary:

> When multiplying two polynomials, each term of one polynomial must be multiplied by each term of the other polynomial. We may also multiply a polynomial by a polynomial using a vertical procedure.

Example 7. Multiply $(3x + 2)(2x^2 - 5x + 2)$

$$(3x + 2)(2x^2 - 5x + 2)$$
$$= 3x(2x^2 - 5x + 2) + 2(2x^2 - 5x + 2)$$
$$= 6x^3 - 15x^2 + 6x + 4x^2 - 10x + 4$$
$$= 6x^3 - 11x^2 - 4x + 4$$

Using vertical procedure, we have,

$$
\begin{array}{r}
2x^2 - 5x + 2 \\
3x + 2 \\ \hline
4x^2 - 10x + 4 \\
6x^3 - 15x^2 + 6x \\ \hline
6x^3 - 11x^2 - 4x + 4
\end{array}
$$

Multiply top polynomial by -3
Multiply top polynomial by $2x^2$ and align like terms
Add like terms in column

Exercises Set 4.5

Multiply.

1. $-5(3x + 8y)$
2. $3y(y^2 + 3y)$
3. $-2a(a^4 + 6)$
4. $-2a^2b(a^2 - 3ab + 5b^2)$
5. $(x - 1)(x + 5)$
6. $(y + 7)(y - 3)$
7. $(4u + 3w)(2u - 3w)$
8. $(2c - 7)(3c + 2)$
9. $(y^2 + 7)(2y^2 - 5)$
10. $(-d^2 - 1)(-2d^2 - 4)$
11. $(2x + 3y)(2x^2 - 3xy + 5y^2)$
12. $(3a - 4b)(9a^2 + 12ab + 16b^2)$

Multiply using a special product formula:

13. $(x - 8)^2$
14. $(y + 7)(y - 7)$
15. $(x + 3y)^2$
16. $(2a + 5)^2$

17. $(r - 6)(r + 6)$ 	18. $(3 - 4a)^2$
19. $(3t - 8u)^2$ 	20. $(4d + 7c)(4d - 7c)$

Answers to Exercises Set 4.5

1. $-15x - 40y$ 	2. $3y^3 + 9y^2$
3. $-2a^5 - 12a$ 	4. $-2a^4b + 6a^3b^2 - 10a^2b^3$
5. $x^2 + 4x - 5$ 	6. $y^2 + 4y - 21$
7. $8u^2 - 6uw - 9w^2$ 	8. $6c^2 - 17c - 14$
9. $2y^4 + 9y^2 - 35$ 	10. $2d^4 + 6d^2 + 4$
11. $4x^3 - xy^2 + 15y^3$ 	12. $27a^3 - 64b^3$
13. $x^2 - 16x + 64$ 	14. $y^2 - 49$
15. $x^2 + 6xy + 9y^2$ 	16. $4a^2 + 20a + 25$
17. $r^2 - 36$ 	18. $16a^2 - 24a + 9$
19. $9t^2 - 48tu + 64u^2$ 	20. $16d^2 - 49c^2$

Section 4.6   Division of Polynomials

Summary

> **To divide a polynomial by a monomial**, divide each term of the polynomial by the monomial.

Example 1. Divide

a) $\dfrac{8x - 24}{4}$

$$\dfrac{8x - 24}{4} = \dfrac{8x}{4} - \dfrac{24}{4}$$
$$= 2x - 6$$

b) $\dfrac{12y^4 + 18y^3 - 27y - 9}{3y}$

$$\dfrac{12y^4 + 18y^3 - 27y - 9}{3y} = \dfrac{12y^4}{3y} + \dfrac{18y^3}{3y} - \dfrac{27y}{3y} - \dfrac{9}{3y}$$

$$= 4y^3 + 6y^2 - 9 - \dfrac{3}{y}$$

Note: A common student error is to fail to divide each term of the polynomial by the monomial.

$$\dfrac{4x + 6}{2} \neq 2x + 6$$

Summary:

> **Dividing a Polynomial by a Binomial:**
>
> When dividing a polynomial by a binomial, use the long division algorithm.

Example 2. Divide $\dfrac{x^2 - 3x - 10}{x + 2}$

$\dfrac{x^2 - 3x - 10}{x + 2}$ → dividend
→ divisor

Write as: $x + 2 \,\overline{\smash{\big)}\, x^2 - 3x - 10}$

Divide x^2 (the first term in the dividend) by x (the first term in the divisor).

$\dfrac{x^2}{x} = x$ (quotient)

Place the quotient, x, above the like term containing x in the dividend.

$$x + 2 \,\overline{\smash{\big)}\, x^2 - 3x - 10}^{\;\;x}$$

Next, multiply x by $x + 2$ as you would do in long division and place the product under their like terms.

$x(x + 2) = x^2 + 2x$

$$\begin{array}{r} x \\ x + 2 \,\overline{\smash{\big)}\, x^2 - 3x - 10} \\ \underline{x^2 + 2x} \end{array}$$

Now, subtract $x^2 + 2x$ from $x^2 - 3x$. Change the signs of the terms being subtracted and add like terms.

$$\begin{array}{r} x \\ x + 2 \overline{\smash{)}\, x^2 - 3x - 10} \\ \underline{-x^2 - 2x} \\ -5x \end{array}$$

Now, bring down -10, the next term in the dividend.

$$\begin{array}{r} x \\ x + 2 \overline{\smash{)}\, x^2 - 3x - 10} \\ \underline{x^2 + 2x} \\ -5x - 10 \end{array}$$

Now, divide the first term at the bottom, by x, the first term in the divisor.

$$\frac{-5x}{x} = -5$$

Write the -5 in the quotient above the constant in the dividend.

$$\begin{array}{r} x - 5 \\ x + 2 \overline{\smash{)}\, x^2 - 3x - 10} \\ \underline{x^2 + 2x} \\ -5x - 10 \end{array}$$

Multiply the x + 2 by the -5 and place the terms of the product under their like terms.
-5(x + 2) = -5x - 10

$$\begin{array}{r} x - 5 \\ x + 2 \overline{\smash{)}\, x^2 - 3x - 10} \\ \underline{x^2 + 2x} \\ -5x - 10 \\ \underline{-5x - 10} \end{array}$$

Subtract (Note there is no remainder), so

$$\frac{x^2 - 3x - 10}{x + 2} = x - 5$$

Summary:

> **To Check Division of Polynomials**
>
> (Divisor x quotient) + remainder = dividend.

Example 3. Divide $\dfrac{2x^2 + 3x - 5}{2x - 3}$

$$
\begin{array}{r}
x + 3 \\
2x - 3 \overline{\smash{)}\, 2x^2 + 3x - 5} \\
\underline{-2x^2 + 3x} \\
6x - 5 \\
\underline{-6x + 9} \\
4
\end{array}
$$

Thus, $\dfrac{2x^2 + 3x - 5}{2x - 3} = x + 3 + \dfrac{4}{2x - 3}$

Check:
$$
\begin{aligned}
(2x - 3)(x + 3) + 4 &= 2x^2 + 3x - 5 \\
2x^2 + 6x - 3x - 9 + 4 &= 2x^2 + 3x - 5 \\
2x^2 + 3x - 5 &= 2x^2 + 3x - 5 \quad \text{True}
\end{aligned}
$$

Example 4. Divide $18 - 4x + 2x^3$ by $x + 4$

First, rewrite the dividend in descending order to get

$(2x^3 - 4x + 18) \div (x + 4)$

Since there is no x^2 term in the dividend, we will add $0x^2$ to help align the terms.

$$
\begin{array}{r}
2x^2 - 8x + 28 \\
x + 4 \overline{\smash{)}\, 2x^3 + 0x^2 - 4x + 18} \\
\underline{-2x^3 - 8x^2 } \\
-8x^2 - 4x \\
\underline{8x^2 + 32x } \\
28x + 18 \\
\underline{-28x - 112} \\
-94
\end{array}
$$

Thus, $\dfrac{2x^3 - 4x + 18}{x + 4} = 2x^2 - 8x + 28 + \dfrac{-94}{x + 4}$

Exercises Set 4.6

Divide.

1. $\dfrac{10a - 25}{5}$

2. $\dfrac{6y^2 + 4y}{y}$

3. $\dfrac{3x^2 - 6x}{-3x}$

4. $\dfrac{x^6 - 3x^4 - x^2}{x^2}$

5. $\dfrac{8x^2y^2 - 24xy}{8xy}$

6. $\dfrac{3x^2 - 2x + 1}{x}$

7. $\dfrac{2y^2 - 6y + 9}{y}$

8. $\dfrac{9x^2y + 6xy - 3xy^2}{xy}$

9. $\dfrac{x^2 + 10x + 25}{x + 5}$

10. $\dfrac{y^2 + 2y - 35}{y + 7}$

11. $\dfrac{2y^2 + 7}{y - 3}$

12. $\dfrac{6y^2 + 2y}{2y + 4}$

13. $\dfrac{b^2 - 8b - 9}{b - 3}$

14. $\dfrac{10y^2 + 21y + 10}{2y + 3}$

15. $\dfrac{6a^2 + 25a + 24}{3a - 1}$

16. $\dfrac{4a^3 + 8a^2 + 5a + 9}{2a + 3}$

17. $\dfrac{6a^3 + 5a^2 + 5}{3a + 2}$

18. $\dfrac{4b^3 - b - 5}{2b + 3}$

19. $\dfrac{6x + 2x^3 + 26}{x + 2}$

20. $\dfrac{x^3 - 2x - 4}{x - 2}$

Answers to Exercises Set 4.6

1. $2a - 5$
2. $6y + 4$
3. $-x + 2$
4. $x^4 - 3x^2 - 1$
5. $xy - 3$
6. $3x - 2 + \dfrac{1}{x}$
7. $2y - 6 + \dfrac{9}{y}$
8. $9x + 6 - 3y$
9. $x + 5$
10. $y - 5$
11. $2y + 6 + \dfrac{25}{y - 3}$
12. $3y - 5 + \dfrac{20}{2y + 4}$
13. $b - 5 - \dfrac{24}{b - 3}$
14. $5y + 3 + \dfrac{1}{2y + 3}$
15. $2a + 9 + \dfrac{33}{3a - 1}$
16. $2a^2 + a + 1 + \dfrac{6}{2a + 3}$
17. $2a^2 + 3a + 2 + \dfrac{9}{3a - 2}$
18. $2b^2 - 3b + 4 - \dfrac{17}{2b + 3}$
19. $2x^2 - 4x + 14$
20. $x^2 + 2x + 2$

Section 4.7   **Motion and Mixture Problems**

Summary

> **Motion Problems**
>
> Amount = rate · time

Example 1. A swimming pool is being emptied at the rate of 18 gallons per minute. If the pool contained 5400 gallons of water, how long will it take to empty the pool?

Solution. Let x = time in minutes

$$\text{Amount} = \text{rate} \cdot \text{time}$$
$$5400 = 18 \cdot x$$
$$\frac{5400}{18} = x$$
$$300 = x$$

It takes 300 minutes (5 hours) to empty the pool.

Summary

> When the "amount" in the rate formula is "distance", we often refer to the formula as the **distance formula**.
>
> Distance = rate · time or
> d = r · t

Example 2. A trip takes an automobile a total of 8 hours. If the trip is a distance 440 miles, what is the average rate of the car on the trip?

Solution.
Let x = average rate of the car

$$\text{Distance} = \text{rate} \cdot \text{time}$$
$$440 = x \cdot 8$$
$$\frac{440}{80} = x$$
$$55 = x$$

The average rate of the car is 55 miles per hour.

Example 3: Two cars, one traveling 10 mph faster than the second car, start at the same time and travel in opposite directions. In 3 hours, they are 300 miles apart. Find the rate of each car.

Solution: Let x = rate of one car
Let x + 10 = rate of the faster car

	rate	·	time	=	distance
first car	x		3		3x
faster car	x + 10		3		3(x + 10)

$$\begin{aligned}
\text{Distance of first car} + \text{Distance of faster car} &= 300 \\
3x + 3(x+10) &= 300 \\
3x + 3x + 30 &= 300 \\
6x + 30 &= 300 \\
6x + 30 - 30 &= 300 - 30 \\
6x &= 270 \\
\frac{6x}{6} &= \frac{270}{6} \\
x &= 45
\end{aligned}$$

So, first car rate is 45 mph. Faster car rate is x + 10 = 45 + 10 = 55 mph.

Example 4. On a 105 mile trip, a car traveled at an average speed of 45 mph and then reduced the speed to 30 mph for the remainder of the trip. The trip took a total of 150 minutes. For how long did the car traveled at 30 mph?

Solution.

First of all convert 150 minutes to hours

$$150 \text{ minutes} = 150(\text{minutes})\left(\frac{1 \text{ hr.}}{60 \text{ minutes}}\right) = 2.5 \text{ hrs}$$

Let x = travel time at 30 mph
Let 2.5 - x = travel time at 45 mph

	rate	·	time	=	distance
30 mph	30		x		30x
45 mph	45		2.5 - x		45(2.5 - x)

$$\begin{aligned}
(\text{Distance at 30 mph}) + (\text{Distance at 45 mph}) &= (\text{Total distance}) \\
30x + 45(2.5 - x) &= 105 \\
30x + 112.5 - 45x &= 105 \\
-15x + 112.5 &= 105
\end{aligned}$$

$$
\begin{aligned}
-15x + 112.5 - 112.5 &= 105 - 112.5 \\
-15x &= -7.5 \\
\frac{-15x}{-15} &= \frac{-7.5}{-15} \\
x &= 0.5
\end{aligned}
$$

So, the car travels 0.5 hrs (30 minutes) at 30 mph.

Summary

> Any problem in which two or more quantities are combined to produce a different quantity or a single quantity is separated into two or more different quantities may be considered a **mixture problem**.

Example 5. A coffee merchant wants to make 6 lbs of a blend of coffee costing $5 per pound. The blend is made using a $6 grade and a $3 grade of coffee. How many pounds of each of these grades should be used?

Solution: Let x = number of lbs of $6 grade
Let 6 - x = number of lbs of $3 grade

Price	Number of Pounds	Values of Coffee
$6	x	6x
$3	6 - x	3(6 - x)
$5	6	5 · 6

(Value of $6 coffee) + (value of $3 coffee) = (value of $5 coffee)

$$
\begin{aligned}
6x + 3(6 - x) &= 5 \cdot 6 \\
6x + 18 - 3x &= 30 \\
3x + 18 &= 30 \\
3x + 18 - 18 &= 30 - 18 \\
3x &= 12 \\
\frac{3x}{3} &= \frac{12}{3} \\
x &= 4
\end{aligned}
$$

So, 4 lbs at $6 per lb. and 6 - 4 = 2 lbs at $3 per lb.

Example 6. How many gallons of a 20% salt solution must be mixed with 6 gallons of a 30% salt solution to makes a 22% salt solution?

Solution.

Let x = the number of gallons of a 20% salt solution

Solution	Strength	Gallons	Amounts of Pure Salt
20%	0.20	x	0.20x
30%	0.30	6	6(0.30)
22%	0.22	x + 6	0.22(x + 6)

Amount of Pure Salt   +   Amount of Pure Salt   =   Amount of Pure Salt
   in 20% solution   in 30% solution   in 22% solution

```
       0.20x           +        6(0.30)       =   0.22(x + 6)
       0.20x           +          1.8         =   0.22x + 1.32
  0.20x - 0.22x        +          1.8         =   0.22x - 0.22x + 1.32
     - 0.02x           +          1.8         =   1.32
     - 0.02x           +      1.8 - 1.8       =   1.32 - 1.8
     - 0.02x                                  =   -0.48

       -0.02x                                 =   -0.48
       ------                                     ------
       -0.02                                      -0.02

         x                                    =   24
```

So, we must mix 24 gallons of the 20% solution.

Example 7. We have a total of $80,000 to be invested. Part will be invested at 5% simple interest and the rest at 10% simple interest. How much should be invested at each rate to return a total interest of 5000 for one year?

Solution: Let x = amount of invested at 5%
            Let 80000 - x = amount of invested at 10%

Account	Principal	Rate	Time	Interest
5%	x	0.05	1	0.05x
10%	80000 - x	0.10	1	.10(8000 - x

Interest from 5%   +   Interest from 10%   =   Total investment

```
        account             account
        0.05x       +    .10(80000 - x)   =  5000
        0.05x       +    8000 - .10x      =  5000
       -0.05x       +    8000             =  5000
       -0.05x       +    8000 - 8000      =  5000 - 8000
       -0.05x                              = -3000
```

$$\frac{-0.05x}{-0.05} = \frac{-3000}{-0.05}$$

$$x = 60{,}000$$

So, invest $60,000 at 5% rate and $80,000 - $60,000 = $20,000 at 10% rate.

Exercises Set 4.7

Set up an equation that can be used to solve each problem. Solve the equation and answer the question. Use a calculator when you feel it is appropriate.

1. How fast must a plane travel to fly 1200 miles in 3 hours?

2. A person walked a total of 6 miles in $1\frac{1}{2}$ hours. At what rate did the person walk?

3. A patient is to receive 1200 cubic centimeters of an intravenous fluid over a period of 8 hours. What should be the intravenous flow rate?

4. A typist can type 900 words in 12 minutes. At what rate can this typist type?

5. Two cyclists start from the same point and ride in opposite directions. One cyclist ride twice as fast as the other. In three hours, they are 72 miles apart. Find the rate of each of cyclist.

6. A motorboat leaves a harbor and travels at an average rate of 8 mph toward a small island. Two hours later a cabin cruiser leaves the same island and travels at an average speed of 16 mph toward the same island. In how many hours after the cabin cruiser leaves will the cabin cruiser be alongside the motorboat?

7. A family drove to a resort at an average speed of 30 mph and

later returned over the same road at an average speed of 50 mph. Find the distance to the resort if the total driving time was 8 hrs.

8. Running at an average rate of 8 meters/second, a sprinter ran to the end of a track and then jogged back at an average rate of 3 meters/second. The sprinter took $\frac{11}{12}$ of a minute to run to the end of the track and jog back. Find the length of the track.

9. A car traveling at 48 mph overtakes a cyclist who, riding at 12 mph, has a 3 hr head start. How far from the starting point does the car overtake the cyclist?

10. On a 195 mile trip, a car traveled at an average speed of 45 mph and then reduced the speed to 30 mph for the remainder of the trip. The trip took a total of 5 hours. For how long did the car travel at each speed?

11. How many ounces of pure gold that cost $400 per ounce must be mixed with 20 oz. of an alloy that cost $220 an ounce to make an alloy costing $300 per ounce?

12. How many pounds of walnuts that cost $1.60 per pound must be mixed with 18 lbs of cashews that cost $2.50 per pound to make a mixture that costs $1.90 per pound?

13. Find the cost per pound of a mixture of coffee made from 25 lb of coffee that costs $4.82 per pound and 40 lb of coffee that costs $3.00 per pound.

14. A farmer has some cream that is 21% butterfat and some that is 15% butterfat. How many gallons of each must be mixed to produce 60 gallons of cream that is 19% butterfat?

15. A goldsmith has 10 grams of a 50% gold alloy. How many grams of pure gold should be added to the alloy to make a mixture that is 75% gold?

16. A manufacturer mixed a chemical that was 60% fire retardant with 70 lb of a chemical that was 80% fire retardant to make a mixture that is 74% fire retardant. How much of the 60% mixture was used?

17. A total of $5000 is deposited into two simple interest

accounts. On one account, the annual simple interest rate is 7%, while on the second account, the annual simple interest rate is 9%. How much should be invested in each account os that the total interest earned is $410?

18. A club invested a part of $10,000 in a 6.5% simple annual interest account and the remainder in a 9% annual simple interest account. The amount of interest earned for one year was $725. How much was invested in each account?

19. An investment of $4000 is made at an annual simple interest rate of 7%. How much additional money must be invested at an annual simple interest rate of 10% so that the total interest earned is 8% of the total investment?

20. An investment of $3500 is made at an annual simple interest rate of 6%. How much additional money must be invested at an annual simple interest rate of 9% so that the total interest earned is 8 % of the total investment?

Answers to Exercises Set 4.7

1. 400 mph
2. 4 mph
3. 150 cubic centimeters per hour
4. 75 words per minute
5. 8 mph, 16 mph
6. 2 hrs
7. 150 miles
8. 120 meters
9. 48 miles
10. 3 hrs at 45 mph
    2 hrs at 30 mph
11. 16 oz.
12. 36 lbs
13. $3.70 per lb
14. 40 gallons of 21% butterfat
    20 gallons of 15% butterfat
15. 10 grams
16. 30 lbs.
17. $2000 at 7%
    $3000 at 9%
18. $7000 at 6.5%
    $3000 at 9%
19. $2000
20. $7000

Chapter 4 Practice Test

Simplify each of the following expressions:

1. $4y^3 \cdot 3y^5$
2. $(2x^4)^3$
3. $\dfrac{12a^5}{6a^3}$
4. $\left(\dfrac{2t^5s^2}{6t^2s}\right)^3$
5. $(3b^2c^{-4})^{-3}$
6. $\dfrac{18k^7v^{-4}}{3k^9v^2}$

Determine whether each expression is a polynomial. If the polynomial has a specific name, give that name:

7. $x^{-4} + 3x$
8. $2x$
9. $4x^2 + 3$
10. Write the polynomial $-3 + 4x^5 - 6x^3 - 7x$ in descending order, and gives its degree.

Perform the indicated operations:

11. $(2x^2 + 5x - 2) + (6x^2 - 9x + 5)$
12. $(4x^2 - 8x - 5) - (10x^2 + 5x - 11)$
13. $(x^3 + x^2 - 1) - (2x^2 + 5x)$
14. $(3x - 1)(2x + 5)$
15. $(8 - 3x)(7 + 2x)$
16. $5x^2(2x^2 - 5x - 3)$
17. $(3x + 2)(x^2 - 5x + 2)$
18. $\dfrac{25y^2 + 15y + 20}{-5}$
19. $\dfrac{8a^4 + 6a^2 + 4}{2a}$
20. $\dfrac{8x^2 - 30x + 7}{2x - 7}$

21. A bricklayer can lay 200 bricks in 40 minutes. At what rate can he lay bricks per hour?

22. A car travels a distance of 320 miles in a total of 7 hours. If it travels part of the time at 50mph and part of the time

at 40 mph, determine the time spent traveling at the two different rate.

23. A total of $40,000 is to be invested. Part will be invested at 8% annual simple interest rate and the remainder at 5% annual simple interest rate. How much should be invested at each rate to have a total interest returned of $2300?

Answers to Practice Test, Chapter 4

1. $12y^8$
2. $8x^{12}$
3. $2a^2$
4. $\dfrac{t^9 s^3}{27}$
5. $\dfrac{c^{12}}{27b^6}$
6. $\dfrac{6}{k^2 v^6}$
7. Not a polynomial
8. monomial
9. binomial
10. $4x^5 - 6x^3 - 7x - 3$
11. $8x^2 - 3x + 3$
12. $-6x^2 - 13x + 6$
13. $x^3 - x^2 - 5x - 1$
14. $6x^2 + 13x - 5$
15. $-6x^2 - 5x + 56$
16. $10x^4 - 25x^3 + 15x^2$
17. $3x^3 - 13x^2 - 4x + 4$
18. $-5y^2 - 3y - 4$
19. $4a^3 + 3a + \dfrac{2}{a}$
20. $4x - 1$
21. 300 bricks per hr.
22. 4 hrs at 50 mph
    3 hrs at 40 mph
23. $10,000 at 8%
    $30,000 at 5%

Section 5.1   Factoring a monomial from a polynomial

Summary:

> 1. To factor an expression means to write the expression as a product of its factors.
>
> 2. If a · b = c, then a and b are factors of c.
>
> 3. The greatest common factor (GCF) of two or more numbers is the greatest number that divides all the numbers.
>
> 4. To determine the GCF of 2 or more numbers:
> a) Write each number as a product of prime numbers.
>
> b) Determine the prime factors common to all the numbers.
>
> c) Multiply the common factors found in previous step. The product of these factors is the GCF.

Example 1. Find the GCF of 24 and 40. Write each number as a product of prime numbers.

   1. 24 = 2 · 2 · 2 · 3   (Recall 2, 3 and 5
   2. 40 = 2 · 2 · 2 · 5   are prime numbers)
   3. Factors common to both 24 and 40 are 2 · 2 · 2
   4. Multiply these factors: 2 · 2 · 2 = 8
   5. GCF is 8

Example 2. Find the GCF of y^6, y^3, y^5, and y^4.

The GCF is y^3 since y^3 is the highest power of y that divides each term.

Summary

> **Greatest Common Factor of 2 or More Terms**
>
> To find the GCF of 2 or more terms, take each factor the **fewest** number of times it appears in any of the terms.

Example 3. Find the GCF of a^3b^2c, a^2bc^2, ab^3c^3.

The highest power of a common to all 3 terms is a^1 or a. The highest power of b common to all three terms is b^1. The highest power of c common to all 3 terms is c^1 or c. So the GCF of the 3 terms is abc.

Example 4. Find the GCF of $x^3(x + 2)^2$, $x^2(x + 2)$, $x^2(x + 2)^3$.

The smallest power of x that appears in all 3 factors is 2 so x^2 is part of the GCF. The smallest power of (x + 2) that appears is 1 so (x + 1) is part of the GCF. Thus the GCF is $x^2(x + 2)$.

Example 5. Find the GCF of set of terms below.
$9x^3$, $45x^2$, $21x^5$

Since $9 = 3 \cdot 3$, $45 = 3 \cdot 3 \cdot 5$, and $21 = 3 \cdot 7$ the GCF of 9, 45, and 21 is 3. Also the smallest power of x common to all 3 terms is 2 so x^2 is part of the GCF. Thus, the GCF is $3 \cdot x^2$.

Summary

> **To Factor a Monomial from a Polynomial**
>
> 1. Determine the GCF of all terms in the polynomial.
>
> 2. Write each term as the product of the GCF and its other factor.
>
> 3. Use the distributive property to factor out the GCF.
>
> 4. To check the factoring process, multiply the factors using the distributive property.

Example 6.   Factor $8x^2 + 12x$

    1. Since $8 = 2 \cdot 2 \cdot 2$ and $12 = 2 \cdot 2 \cdot 3$, then the GCF of 8 and 12 is $2 \cdot 2 = 4$.

    2. The largest power of x that appears is 1 so x^1 is part of the GCF.

    3. The GCF is thus $4x$.

    4. Write each term as the product of the GCF and its other factor. $8x^2 + 12x = 4x(2x) + 4x(3)$

    5. Factor out the GCF using the distributive property $4x(2x + 3)$.

Example 7.   Factor $8y^4 - 12y^2 + 40y$

    1. Find GCF: Since $8 = 2 \cdot 2 \cdot 2$, $12 = 2 \cdot 2 \cdot 3$, $40 = 2 \cdot 2 \cdot 2 \cdot 5$ the GCF of 8, 12 and 40 is 4.

    2. The GCF of y^4, y^2, and y is y.

    3. The GCF of $8y^4$ $12y^2 + 40y$ is $4y$

    4. $8y^4 - 12y^2 + 40y = 4y(2y^3) + 4y(3y) + 4y(10)$
       $= 4y(2y^3 - 3y + 10)$

    5. Check: $4y(2y^3 - 3y + 10) = 8y^4 - 12y^2 + 40y$

Example 8. Factor $2x(4x - 3) + 3(4x - 3)$

    1. The GCF is $(4x - 3)$
    2. Use the distributive property to factor out the GCF: $2x(\underline{4x - 3}) + 3(\underline{4x - 3}) = (2x + 3)(4x - 3)$
    The answer may also be expressed as $(4x - 3)(2x + 3)$

Example 9. $x(x + 4) - (x + 4)$

    Rewrite: $x(x + 4) - 1(x + 4)$
    The GCF is $(x + 4)$
    Factor out the GCF using distributive property
    $x(\underline{x + 4}) - 1(\underline{x + 4}) = (x - 1)(x + 4)$
    or $(x + 4)(x - 1)$

Example 10. Factor $4y^2 + 7xz + 5z^2$

    The only factor common to all 3 terms is one. The polynomial cannot be factored by the method presented in this section.

Summary:

> All factoring problems can and should be checked by multiplying the factors. The product of all the factors should equal the original expression. Also, the <u>first</u> step in factoring a polynomial by the methods of this chapter will always be to determine if there is a common factor of each term. You should always factor out the GCF first.

Exercise Set 5.1

Write each number as a product of prime numbers:

1. 70
2. 91
3. 68
4. 54
5. 245
6. 175

Find the GCF for the 2 given numbers:

7. 68, 12
8. 72, 90
9. 51, 34
10. 120, 72

Find the GCF for each set of terms:

11. $6x, 4x^2, 8x^3$
12. xy^2, x^2y, x^2y^2
13. $18x^4, 6x^2y, 9y^2$
14. $2y - 3, x(2y - 3)$
15. $-8x^2, -9x^3, 12xy^3$

Factor the GCF from each term in the expression. If an expression cannot be factored, so state.

16. $3x + 3y$
17. $15y - 5$
18. $9y^2 - 6y$
19. $36a^{12} - 24a^6$
20. $y + 3yx^2$
21. $7x + 5y$
22. $(m + n)(m + n) - (m + n)$
23. $52x^2y^2 + 16xy^3 + 26z$
24. $4y(2x + 1) + (2x + 1)$
25. $15x^2 - 16x + 7$

Answers to Exercise Set 5.1

1. $2 \cdot 5 \cdot 7$
2. $13 \cdot 7$
3. $2^2 \cdot 17$
4. $2 \cdot 3^3$
5. $5 \cdot 7^2$
6. $5^2 \cdot 7$
7. 4
8. 18
9. 17
10. 24
11. $2x$
12. xy
13. 3
14. $2y - 3$
15. x
16. $3(x + y)$
17. $5(3y - 1)$
18. $3y(3y - 2)$
19. $12a^6(3a^6 - 2)$
20. $y(1 + 3x^2)$
21. cannot be factored
22. $(m + n)(m + n - 1)$

23. $2(26x^2y^2 + 8xy^3 + 13z)$ 24. $(2x + 1)(4y + 1)$

25. cannot be factored

Section 5.2   Factoring By Grouping

Summary

> **To Factor a Four-Term Polynomial Using Grouping**
>
> 1. Determine if there are any factors common to all four terms. If so, factor the GCF from each of the four terms.
>
> 2. If necessary, arrange the terms so that the first 2 terms have a common factor and the last two have a common factor.
>
> 3. Use the distributive property to factor each group of two terms.
>
> 4. Factor the GCF from the result of step 3.

Example 1.   Factor by grouping: $x^2 + 5x + 2x + 10$

1. Check to see if there is a common factor among the four terms. - NO.
2. Rearrange terms $(x^2 + 5x) + (2x + 10)$
3. $(x^2 + 5x) + 2x + 10) =$
   $x(x + 5) + 2(x + 5)$
4. Factor out the common binomial factor of $x + 5$
   $(x + 2)(x + 5)$

Note: The like terms 5x and 2x of this example could have been combined. But our purpose of this example was to illustrate the method of factoring by grouping so they were not combined.

Example 2.   Factor $3x^2 - 15x + 4x - 20$ by grouping

1. No common factor among all 4 terms
2. Rearrange $(3x^2 - 15x) + (4x - 20)$

Note: 3x is common to first pair and 4 is common to last pair.

3. $3x(x - 5) + 4(x - 5)$
4. Factor out common binomial of $(x - 5)$
   $(3x + 4)(x - 5)$ or
   $(x - 5)(3x + 4)$

Example 3. Factor $5x^2 - 25x - x + 5$ by grouping

1. No common factor among all 4 terms
2. Rearrange: $(5x^2 - 25x) - x + 5$
3. $5x$ is GCF of first pair
   $5x(x - 5) - x + 5$
4. Here, we must factor out a negative one from the second binomial to make it identical with the factor $(x - 5)$ of the first pair:
   $5x(x - 5) - 1(x - 5)$
5. Factor out common binomial $(x - 5)$
   $(5x - 1)(x - 5)$

Example 4. Factor $-x^3 + bx^2 + 3b^2x - 3b^3$ by grouping

1. No common factor among all 4 terms
2. Rearrange: $(-x^3 + bx^2) + (3b^2x - 3b^3)$
3. x^2 is common to first pair and $3b^2$ is GCF of second pair.
4. $x^2(-x + b) + 3b^2(x - b)$
   Here, we must adjust the sign of one of the binomial factors to make it identical with the other binomial term. Factor out a negative one from first binomial: $x^2(-1)(x - b) + 3b^2(x - b)$. Now, factor out common binomial $(x - b)$
   $(-x^2 + 3b^2)(x - b)$ or
   $(3b^2 - x^2)(x - b)$

Example 5. Factor $6x + 5y + xy + 30$

1. No common factor among all 4 terms.
2. Note: No common factors among the first pair.
3. Rearrange: $6x + xy + 5y + 30$
4. $(6x + xy) + (5y + 30)$
5. x is common to first pair and y is common to second pair.
   $x(6 + y) + 5(y + 6) =$
   $x(y + 6) + 5(y + 6)$
6. Factor out common binomial $(y + 6)$
   $(x + 5)(y + 6)$

Example 6. Factor $2x^3 - 5x^2 - 6x^2 + 15x$ by grouping

1. Notice that x is the GCF.
2. $x(2x^2 - 5x - 6x + 15)$
3. $x[(2x^2 - 5x) - 6x + 15] =$
   $x[x(2x - 5) - 3(2x - 5)] =$
   $x[(x - 3)(2x - 5)]$

Exercise Set 5.2

Factor by grouping.

1. $x^3 - x^2 + 7x - 7$
2. $ax - ay + bx - by$
3. $x^2 + bx - ax - ab$
4. $y^3 + 2y^2 - 4y - 8$
5. $x^2 - 3x + 2xy - 6y$
6. $x^2 - 2b + 3ax - 6ab$
7. $2t^2 - 10t + 3t - 15$
8. $x^3 + 3x^2 + 2x$
9. $y^2 - yb + ya - ab$
10. $18m - 12m^2 + 24m^3 - 16m^4$
11. $x^2 - x - x + 1$
12. $y^2 - 3y - 3y + 9$
13. $ax + ya - bx - by$
14. $2y^2 - 4xy + 8xy - 16x^2$
15. $a^3 - 2a^2b + 2b^2a - 4b^3$
16. $2cx + vx + 4cy + 2yv$
17. $36x^3 - 18x^2 - 84x + 42$
18. $y^2 - 4by + 3ay - 12ab$
19. $xy - x + 5y - 5$
20. $ca - 2b + 2a - cb$

Answers to exercise set 5.2

1. $(x^2 + 7)(x - 1)$
2. $(x - y)(a + b)$
3. $(x + b)(x - a)$
4. $(y^2 - 4)(y + 2)$
5. $(x + 2y)(x - 3)$
6. $(x + 3a)(x - 2b)$
7. $(2t + 3)(t - 5)$
8. $x(x + 2)(x + 1)$
9. $(y + a)(y - b)$
10. $2m(3 + 4m^2)(3 - 2m)$
11. $(x - 1)(x - 1) = (x - 1)^2$
12. $(y - 3)((y - 3)$
13. $(a - b)(x + y)$
14. $(2y + 8x)(y - 2x)$
15. $(a^2 + 2b^2)(a - 2b)$
16. $(2c + v)(x + 2y)$
17. $6(3x^2 - 7)(3x - 1)$
18. $(y - 4b)(y + 3a)$

19. $(x + 5)(y - 1)$ 20. $(x + 2)(a - b)$

Section 5.3   Factoring Trinomials with a = 1

Summary

> To Factor Trinomials of the form $ax^2 + bx + x$, where $a = 1$:
>
> 1. Find two numbers whose product equals the constant c and whose sum equals the coefficient b.
>
> 2. Using the 2 numbers found in step one, the factored form of $ax^2 + bx + c$ will be (x + one number)(x + second number).
>
> 3. Hint: If the constant c is positive, then both factors must have the same sign; either both are positive or both are negative.
> If the constant c is negative, then the two factors will have opposite signs.

Example 1.   Factor $x^2 - 7x + 12$

    1. Since c is positive both factors must have the same sign, negative, since the coefficient of x is -7.
    2. Make a chart:

Factors of +12	Associated Sums
-1, -12	-13
-2, -6	-8
-3, -4	-7

       Thus -3 and -4 are the required factors.
    3. The factorization is (x - 3)(x - 4)
    4. Check by FOIL. $(x - 3)(x - 4) = x^2 - 4x - 3x + 12$
                                            $= x^2 - 7x + 12$

Example 2. Factor $x^2 + 8x - 20$

1. Since c is negative, required factors must have opposite signs.
2. Factors of -20   Associated Sums
   -1, 20   19
   1, -20   -19
   2, -10   -8
   -2, 10   8
   (We can stop here, since we have found the desired factors)

   -2 and 10 are the required numbers.

3. $x^2 + 8x - 20 = (x + 10)(x - 2)$
4. Check $(x + 10)(x - 2) = x^2 - 2x + 10x - 20$
   $= x^2 + 8x - 20$

Example 3. $y^2 + 11y - 30$

1. Since c is negative, factors must have opposite signs.
2. Factors of -30   Associated Sums
   -1, 30   29
   1, -30   -29
   2, -15   -13
   -2, 15   13
   3, -10   -7
   -3, 10   7
   5, -6   -1
   -5, 6   1

Since we cannot find any factors of -30 whose sum is 11, we conclude that the trinomial cannot be factored and is prime.

Example 4. Factor: $x^2 - 13xy - 48y^2$

This example has 2 variables but the procedure is essentially the same. Just remember that the product of the last terms of the binomials must be $-48y^2$.

1. Since -48 is negative, the factors must have opposite signs.
2. Factors of -48   Associated Sums
   -1, 48   47
   1, -48   -47
   2, -24   -22
   -2, 24   22
   3, -16   -13

3. Thus, 3 and -16 are the desired factors.

4. Write the answer $(x + 3y)(x - 16y)$
5. Check $(x + 3y)(x - 16y) =$
   $x^2 - 16xy + 3xy - 48y^2 =$
   $x^2 - 13xy - 48y^2$

Example 5: Factor $x^2 - 11x - 60$ using trial and error.

1. Since the last number, -60, is negative, the factors will have opposite signs.

Factors of -60	Possible Factors
-60, 1	$(x - 60)(x + 1) = x^2 - 59x - 60$
-30, 2	$(x - 30)(x + 2) = x^2 - 28x - 60$
-20, 3	$(x - 20)(x + 3) = x^2 - 17x - 60$
-15, 4	$(x - 15)(x + 4) = x^2 - 11x - 60$

   So $x^2 - 11x - 60 = (x - 15)(x + 4)$

Example 6: Factor $2x^2 + 8x - 10$ using any method.

1. Since $a \neq 1$, we check for a common factor
2. 2 is the GCF. So we factor it out
   $2(x^2 + 4x - 5)$
3.
Factors of -5	Associated Sums
-5, 1	-4
5, -1	4

4. $2(x^2 + 4x - 5) = 2(x + 5)(x - 1)$

Example 7: Factor $2x^3 - 12x^2 + 10x$

1. Factor out the GCF of $2x$:
   $2x(x^2 - 6x + 5)$
2.
Factors of 5	Associated Sums
1, 5	6
-1, -5	-6

3. $2x(x - 1)(x - 5)$

Exercises Set 5.3

Factor each expression if possible. If not possible, then so state.

1. $y^2 + 3y + 2$
2. $y^2 - 2y - 15$
3. $b^2 + 4b - 21$
4. $x^2 + 2x - 35$
5. $a^2 + 6a + 9$
6. $x^2 - 8x + 12$

7. $c^2 + cx - 56x^2$
8. $x^2 - 9xy + 20y^2$
9. $m^2 + 4m + 6$
10. $-4y^2 + 4y + 24$
11. $b^2 - 15bz + 36z^2$
12. $4x^2 + 12x - 16$
13. $x^2 - 6xy + 8y^2$
14. $x^3 + 11x^2 - 42x$
15. $2x^3 - 4x^2 - 30x$
16. $x^2 - 17x - 60$
17. $3x^2 - 6x - 24$
18. $x^2 - 14x - 51$
19. $x^2 + 21x + 68$
20. $x^2 - 26x + 69$
21. $x^2 + 7x + 24$
22. $x^2 - 18x - 40$
23. $x^2 + 38x + 72$
24. $x^2 - 28xy - 60y^2$
25. $2x^3 + 6x^2 - 56x$

Answers to Exercise Set 5.3

1. $(y + 2)(y + 1)$
2. $(y - 5)(y + 3)$
3. $(b + 7)(b - 3)$
4. $(x + 7)(x - 5)$
5. $(a + 3)(a + 3)$
6. $(x - 6)(x - 2)$
7. $(c + 8x)(c - 7x)$
8. $(x - 4y)(x - 5y)$
9. Does not factor
10. $-4(y - 3)(y + 2)$
11. $(b - 3z)(b - 12z)$
12. $4(x + 4)(x - 1)$
13. $(x - 4y)(x - 2y)$
14. $x(x + 14y)(x - 3)$
15. $2x(x - 5)(x + 3)$
16. $(x - 20)(x + 3)$
17. $3(x - 4)(x + 2)$
18. $(x - 17)(x + 3)$
19. $(x + 17)(x + 4)$
20. $(x - 23)(x - 3)$
21. Does not factor
22. $(x - 20)(x + 2)$
23. $(x + 36)(x + 2)$
24. $(x - 30y)(x + 2y)$
25. $2x(x + 7)(x - 4)$

Section 5.4   **Factoring trinomials with a ≠ 1**

Summary

> One method of factoring trinomials of form $ax^2 + bx + c$, $a \neq 1$, is the trial and error method.
>
> 1. Determine if there are any common factors to all three terms. If so, factor it out.
>
> 2. Write all pairs of factors of the coefficient of the squared term, a.
>
> 3. Write all pairs of factors of the constant term, c.
>
> 4. Try combinations of these factors until the correct middle term, bx, is found.

Example 1:   Factor $3x^2 + 8x + 4$ using trial and error.

Factors of 4	Possible factors of Trinomial	Sum of the products of the outer and inner terms
1(4)	(3x + 1)(x + 4)	12x + x = 13x
4(1)	(3x + 4)(x + 1)	3x + 4x = 7x
2(2)	(3x + 2)(x + 2)	6x + 2x = 8x

Since (3x + 2)(x + 2) yields the correct middle term, they are the correct factors.

Hint: Since the sign of 4, the constant term, is positive then the factors of 4 must have the same sign. But the sign of the coefficient of the middle term, 8x, is positive, so it was only necessary to consider the positive factors of 4.

Example 2. Factor $2x^2 - 15x - 8$ using trial and error.

Factors of -8	Possible factors of Trinomial	Sum of the products of the outer and inner terms
-8(1)	$(2x - 8)(x + 1)$	-6x
-4(2)	$(2x - 4)(x + 2)$	0x
8(-1)	$(2x + 8)(x - 1)$	6x
4(-2)	$(2x + 4)(x - 2)$	0x
(-1)(8)	$(2x - 1)(x + 8)$	15x
(1)(-8)	$(2x + 1)(x - 8)$	-15x

Thus, our desired factors are $(2x + 1)(x - 8)$ since the sum of the products of the outer and inner terms is -15x.

Example 3. Factor $6x^3 + 5x^2 - 6x$ by trial and error.

1. Remove common x factor: $x(6x^2 + 5x - 6)$

2.

Factors of -6	Possible factors of trinomial $6x^2 + 5x - 6$	Sum of products of outer and inner terms
-6(1)	$(2x - 6)(3x + 1)$	-16x
-3(2)	$(2x - 3)(3x + 2)$	-5x
6(-1)	$(2x + 6)(3x - 1)$	16x
3(-2)	$(2x + 3)(3x - 2)$	5x

Since $(2x + 3)(3x - 2)$ yields the correct sum of the products of the outer and inner terms, this is the correct factorization.

Summary

> A second method for factoring $ax^2 + bx + c$, $a \neq 1$, is the method of grouping.
>
> 1. Factor out the GCF of the three terms.
>
> 2. Find two numbers whose product is equal to ac and whose sum is equal to b.
>
> 3. Rewrite the middle term, bx, as the sum or difference of two terms using the numbers found in step 2.
>
> 4. Factor by grouping

Example 4. Factor $3x^2 + 17x + 10$

    1. GCF is 1
    2. $a = 3 \quad b = 17 \quad c = 10$
    3. set up a chart:

Factors of ac (30)	Associated Sums
1(30)	30 + 1 = 31
2(15)	2 + 15 = 17

       Since $2 \cdot 15 = 30$ and $2 + 15 = 17$, the are the desired numbers.

    4. Rewrite $3x^2 + 2x + 15x + 10$
    5. Factor by grouping $(3x^2 + 2x) + (15x + 10)$
                               $x(3x + 2) + 5(3x + 2)$
                                 $(x + 5)(3x + 2)$

Example 5. Factor $5x^2 - 11x - 12$

    1. GCF is 1
    2. $a = 5, b = -11, c = -12$

Factors of ac = -60	Associated Sums
-1(60)	-1 + 60 = 59
-2(30)	-2 + 30 = 28
-3(20)	-3 + 20 = 17
-4(15)	-4 + 15 = 11
4(-15)	4 + (-15) = -11

Thus 4, -15 are the required numbers.

3. Rewrite $5x^2 + 4x - 15x - 12$
4. Factor by grouping $(5x^2 + 4x) - 15x - 12$
$$x(5x + 4) - 3(5x + 4)$$
$$(x - 3)(5x + 4)$$

Notice: In this step, it was necessary to factor out a -3 rather than a 3 from the last two terms.

Example 6. Factor $18x^3 - 21x^2 - 9x$

1. GCF is $3x$. Factor it out.
$3x(6x^2 - 7x - 3)$

2. Factor $6x^2 - 7x - 3$; $a = 6$, $b = -7$, $c = -3$

3. <u>Factors of ac = -18</u>   <u>Associated Sums</u>
    -18(2)   -18 + 2 = -16
    -9(2)   -9 + 2 = -7

4. -9 and 2 are desired numbers
Rewrite $6x^2 - 7x - 3 = 6x^2 - 9x + 2x - 3$

5. Factor by grouping:
$(6x^2 - 9x) + (2x - 3)$
$3x(2x - 3) + (2x - 3)(1)$ (is implied)
$(3x + 1)(2x - 3)$
6. Final answer: $3x(3x + 1)(2x - 3)$

Example 7. Factor $3x^2 - 7x - 8$

1. GCF is 1

2. $a = 3$, $b = -7$, $c = -8$

3. <u>Factors of ac = -24</u>   <u>Associated Sums</u>
    -24(1)   -23
    -12(2)   -10
    -8(3)   -5
    -6(4)   -2
    -4(6)   2
    -2(12)   10

We can see it is not possible to find factors of -24 whose sum is -7 so we conclude this trinomial is not factorable.

Exercise Set 5.4

Factor completely. If an expression cannot be factored, so state.

1. $6x^2 - 11x - 3$
2. $2x^2 + 5x - 7$
3. $3x^2 - 5x + 2$
4. $10x^2 + x - 3$
5. $6a^2 + 5ab - 6b^2$
6. $6x^2 - 23x + 20$
7. $6x^2 + x - 1$
8. $4x^2 - 5x + 1$
9. $12u^2 - 11uv + 2v^2$
10. $42x^2 + 4xy - 6y^2$
11. $10m^2 - 24m + 8$
12. $6b^2 - 3b - 84$
13. $5x^2 + 2x + 7$
14. $8x^2 + 13x - 6$
15. $15x^2 - 19x + 6$
16. $10x^3 + 17x^2 + 3x$
17. $9x^2 - 6x + 1$
18. $25x^2 - 20x + 4$
19. $56y^2 + 5yz - 6z^2$
20. $2x^2 - 7xy + 3y^2$

Answers to Exercise Set 5.4

Factor completely. If an expression cannot be factored, so state.

1. Cannot be factored
2. $(x - 1)(2x + 7)$
3. $(x - 1)(3x - 2)$
4. $(5x + 3)(2x - 1)$
5. $(2a + 3b)(3a - 2b)$
6. $(3x - 4)(2x - 5)$
7. $(2x + 1)(3x - 1)$
8. $(x - 1)(4x - 1)$
9. $(3u - 2v)(4u - v)$
10. $2(3x - 4)(7x + 3y)$
11. $2(5m - 2)(m - 2)$
12. $3(2b + 7)(b - 4)$
13. Cannot be factored
14. $(8x - 3)(x + 2)$
15. $(5x - 3)(3x - 2)$
16. $x(5x + 1)(2x + 3)$
17. $(3x - 1)^2$
18. $(5x - 2)^2$
19. $(8y + 3z)(7y - 2z)$
20. $(2x - y)(x - 3y)$

Section 5.5 Special Factoring Formulas and a general review of factoring

Summary

Difference of Two Squares

$a^2 - b^2 = (a - b)(a + b)$

Example 1. Factor each of the following using the difference of two squares formula.

a) $x^2 - 16 = x^2 - 4^2$ Replace a with x, b with y
$\qquad a^2 - b^2 = (a + b)(a - b)$
$\qquad x^2 - 4^2 = (x + 4)(x - 4)$

b) $9 - 16x^2 = 3^2 - (4y)^2$ $a = 3, b = 4y$
$\qquad a^2 - b^2 = (a + b)(a - b)$
$\qquad 3^2 - (4y)^2 = (3 + 4y)(3 - 4y)$

c) $49x^2 - 9y^2 = (7x)^2 - (3y)^2$; $a = 7x$ and $b = 3y$
$\qquad a^2 - b^2 = (a + b)(a - b)$
$\qquad (7x)^2 - (3y)^2 = (7x + 3y)(7x - 3y)$

d) $9y^4 - 16x^4 = (3y^2)^2 - (4x^2)^2$; $a = 3y^2; b = 4x^2$
$\qquad a^2 - b^2 = (a + b) \cdot (a - b)$
$\qquad (3y^2)^2 - (4x^2)^2 = (3y^2 + 4x^2)(3y^2 - 4x^2)$

e) $8x^2 - 72v^2$

   1. First, remove the GCF
   $8(x^2 - 9y^2) = 8((x)^2 - (3y)^2)$; $a = x, b = 3y$
   $\qquad = 8(x - 3y)(x + 3y)$

f) $x^4 - 1 = (x^2)^2 - 1^2$
$\qquad a^2 - b^2 = (a - b)(a + b)$
$\qquad (x^2)^2 - 1^2 = (x^2 - 1)(x^2 + 1)$

We can factor x^2 as $(x + 1)(x - 1)$ so our final answer is:
$x^4 - 1 = (x + 1)(x - 1)(x^2 + 1)$

Remember: Always factor <u>completely</u>.

It is helpful to recal numbers which are perfect squares. The first few perfect squares are: 1, 4, 9, 16, 25, 36, 49, 64, 81, 100, 121, 144, 169, 196, 225 and so on.
Also, any variable raised to an even power will be a perfect square.

$$x^2, \quad x^4 = (x^2)^2 \quad x^6 = (x^3)^2 \quad x^8 = (x^4)^2, \text{ and so on.}$$

Example 2. Factor $144x^6 - 121y^4 =$
$$(12x^3)^2 - (11y^2)^2 =$$
$$a^2 - b^2 = (a - b)(a + b)$$
$$(12x^3)^2 - (11y^2)^2 = (12x^3 - 11y^2)(12x^3 + 11y^2)$$

Example 3. Factor $x^2 + 9$

This is a sum of 2 perfect squares - not a difference of 2 perfect squares. It cannot be factored further using the set of real number.

Summary

Sum of 2 cubes

$a^3 + b^3 = (a + b)(a^2 - ab + b^2)$

Difference of 2 cubes

$a^3 - b^3 = (a - b)(a^2 + ab + b^2)$

Example 4. Factor, using the sum of 2 cubes formula or the difference of 2 cubes formulas.

a) $x^3 + 8 = x^3 + 2^3$ Here $x = a, 2 = b$
$$a^3 + b^3 = (a + b)(a^2 - ab + b^2)$$
$$x^3 + 2^3 = (x + 2)(x^2 - 2x + 4)$$

b) $m^3 - 64 = m^3 - 4^3$ $m = a$ and $4 = b$
$$a^3 - b^3 = (a - b)(a^2 + ab + b^2)$$
$$m^3 - 4^3 = (m - 4)(m^2 + 4m + 16)$$

c) $x^5 - 27x^2$

1. Remove GCF of x^2: $x^2(x^3 - 27)$
2. Factor $x^3 - 27 = x^3 - 3^3$; $x = a, 3 = b$
$$a^3 - b^3 = (a - b)(a^2 + ab + b^2)$$
so $x^3 - 3^3 = (x - 3)(x^2 + 3x + 9)$
3. Final answer. $x^5 - 27x^2 = x^2(x - 3)(x^2 + 3x + 9)$

It is helpful to memorize numbers which are perfect cubes: 1, 8, 27, 64, 125, 216, 343, 512, 729, 1000, ...
Also, variables which are raised to exponents which are

multiples of 3 are perfect cubes:

$$x^3, \quad x^6 = (x^2)^3, \quad x^9 = (x^3)^3, \quad x^{12} = (x^4)^3, \ldots$$

d) $8x^6 + 125 = (2x^2)^3 + 5^3$
$a^3 + b^3 = (a + b)(a^2 - ab + b^2)$
$(2x^2)^3 + 5^3 = (2x^2 + 5)(4x^4 - 10x^2 + 25)$

Summary

> **General Procedure to Factor a Polynomial**
>
> 1. If all the terms of the polynomial have a GCF other than 1, **FACTOR IT OUT**.
>
> 2. If the polynomial has **2 terms**, determine if it is a **difference of 2 squares**, or a **sum or difference of 2 cubes**. If so, factor it using the appropriate formula.
>
> 3. If a polynomial has **three terms**, factor it using either **trial and error** or by **grouping**.
>
> 4. If the polynomial has **more than 3 terms** try **factoring by grouping**.
>
> 5. As a final step, examine your factored polynomial to see if the terms in any factors have a common factor. If you find a common factor, factor it out at this point.

Example 5. Factor the following polynomials completely:

a) $3x^2 - 9x - 12$

    1. Remove GCF of 3. $3(x^2 - 3x - 4)$
    2. Factor trinomial $3(x - 4)(x + 1)$

b) $4ya^2 - 36y$

1. Remove GCF of 4y. $4y(a^2 - 9)$
2. Factor $a^2 - 9 = a^2 - 3^2$ as a difference of 2 squares using
   $a^2 - b^2 = (a - b)(a + b)$
   $a^2 - 3^2 = (a - 3)(a + 3)$
   So $4ya^2 - 36y = 4y(a - 3)(a + 3)$

c) Factor $xz + 2x + yz + 2y$

Since there are four terms, try grouping:
$(xz + 2x) + (yz + 2y) =$
$x(z + 2) + y(z + 2) =$
$(x + y)(z + 2)$

d) Factor $3x^2 - 2x - 16$ (a trinomial)

1. Multiply $ac = 3(-16) = -48$
2. We need factors of -48 whose sum is -2. They are -8 and 6
3. Rewrite $3x^2 - 8x + 6x - 16$
4. Use grouping. $3x^2 - 8x + 6x - 16$
   $x(3x - 8) + 2(3x - 8)$
   $(x + 2)(3x - 8)$

Exercise Set 5.5

Factor each polynomial completely.

1. $x^2 - 64$
2. $4x^2 - 9$
3. $y^4 - 16$
4. $9 - 16x^2y^4$
5. $64x^2 - 9y^2$
6. $4x^2 - 121$
7. $100 - 4y^2$
8. $x^3 + 1$
9. $a^3 + 27b^3$
10. $64 - y^3$
11. $27 - 8x^3$
12. $1 + 27y^3$
13. $125x^3 + 8z^6$
14. $w^2 + 18w + 45$
15. $9x^2 + 18x$
16. $3x^2 - 6x - 72$
17. $4x^3 + 32x^2y + 60xy^2$
18. $ab - a + b - 1$
19. $x^2 - 7x + 7x - 49$
20. $5x^2 + 20xy + 3xy + 12y^2$
21. $4x^4 - 9y^4$
22. $27 - 8y^3$

23. $16 - 25y^2$ 24. $x^2 - 2xy - 15y^2$

25. $55p^3 - 20p^2$

Answers to exercise set 5.5

1. $(x + 8)(x - 8)$ 2. $(2x + 3)(2x - 3)$
3. $(y + 2)(y - 2)(y^2 + 4)$ 4. $(3 - 4xy^2)(3 + 4xy^2)$
5. $(8x - 3y)(8x + 3y)$ 6. $(2x - 11)(2x + 11)$
7. $4(5 - y)(5 + y)$ 8. $(x + 1)(x^2 - x + 1)$
9. $(a + 3b)(a^2 - 3ab + ab^2)$ 10. $(4 - y)(16 + 4y + y^2)$
11. $(3 - 2x)(9 + 6x + 4x^2)$ 12. $(1 + 3y)(1 - 3y + 9y^2)$
13. $(5x + 2z^2)(25x^2 - 10xz^2 + 4z^4)$ 14. $(w + 15)(w + 3)$
15. $9x(x + 2)$ 16. $3(x - 6)(x + 4)$
17. $4x(x + 5y)(x + 3y)$ 18. $(a + 1)(b - 1)$
19. $(x + 7)(x - 7)$ 20. $(5x + 3y)(x + 4y)$
21. $(2x^2 + 3y^2)(2x^2 - 3y^2)$ 22. $(3 - 2y)(9 + 6y + 4y^2)$
23. $(4 + 5y)(4 - 5y)$ 24. $(x + 3y)(x - 5y)$
25. $5p^2(11p - 4)$

Section 5.6 Solving Quadratic Equations Using Factoring

Summary

> Quadratic equations have the form $ax^2 + bx + c = 0$, where a, b, and c are real numbers, $a \neq 0$.

Example 1 $3x^2 + 10x + 8 = 0$ is an example of a quadratic equation. Notice that the left side of the equation is written in descending powers of x and the right side of the equation is zero.

Summary

> 1. Some quadratic equations can be solved by factoring. Other quadratic equations can be solved using two additional techniques discussed in a later chapter.
>
> 2. The **Zero-factor Property** is used to solve a quadratic equation by factoring.
>
> 3. **Zero Factor Property**
> If $ab = 0$ then $a = 0$ or $b = 0$
> (In other words, if the product of two algebraic factors is zero, then **at least one of them** must be 0.

Example 2. Solve $(x - 3)(x + 2) = 0$
Since the product of two factors is zero, by the zero factor property, at least one of the factors is zero.

  1. Set each factor to zero and solve.
     $(x - 3) = 0$
     $x = 3$
     $(x + 2) = 0$
     $x = -2$
     Thus, the solution set is $\{3, -2\}$

Example 3. Solve $(2x + 1)(3x - 5) = 0$

    1. Use the zero-factor property.

        $2x + 1 = 0$ or $3x - 5 = 0$

    2. $\quad 2x + 1 = 0 \quad$ or $\quad 3x - 5 = 0$
$$2x = -1 \qquad\qquad 3x = 5$$
$$x = -\frac{1}{2} \quad \text{or} \quad x = \frac{5}{3}$$

    Solutions are $-\frac{1}{2}$ and $\frac{5}{3}$

Summary

> **To solve a quadratic equation by factoring**
>
> 1. Write the equation in standard form with the squared term positive. This will result in one side of the equation being 0.
>
> 2. Factor the non-zero side of the equation.
>
> 3. Set each factor **containing a variable** equal to zero and solve each equation.
>
> 4. Check each solution found in step 3 in the original equation.

Example 4. Solve $x^2 - 4x = 0$

    1. Set one side equal to 0 (This is done)
    2. Factor non-zero side: $x(x - 4) = 0$
    3. Set each factor equal to zero:
       $x = 0 \quad$ or $\quad x - 4 = 0$
       $x = 0 \quad$ or $\quad x = 4$
       Solutions are $x = 0$ and $x = 4$
    4. Check: if $x = 0$, then $0^2 - 4(0) = 0$
                if $x = 4$, then $4^2 - 4(4) = 0$

Example 5. Solve $x^2 - 11x = 60$

1. Subtract 60 from each side to make the right side equal to 0: $x^2 - 11x - 60 = 0$
2. Factor non-zero side:
   (Factors of -60 which sum to -11 are -15 and 4)
   $(x - 15)(x + 4) = 0$
3. Set each factor equal to zero:
   $x - 15 = 0$ or $x + 4 = 0$
   $x = 15$ $x = -4$
4. Check: if $x = 15$ then $15^2 - 11(15) - 60 = 0$
                               $225 - 165 - 60 = 0$
            if $x = -4$ then $(-4)^2 - 11(-4) - 60 = 0$
                               $16 + 44 - 60 = 0$

Example 6. Solve $4x^2 = 40x - 96$

1. Set right side equal to zero by adding $-40x + 96$ to both sides. This will result in the coefficient of the squared term being positive.
   $4x^2 - 40x + 96 = 0$

2. Now, factor out the GCF of 4:
   $4(x^2 - 10x + 24) = 0$
   Factor $x^2 - 10x + 24$ as $(x - 6)(x - 4)$
   So we now have $4(x - 6)(x - 4) = 0$

3. Set each factor **containing a variable** equal to zero and solve the equations.

   Thus, $x - 6 = 0$ or $x - 4 = 0$
         $x = 6$ or $x = 4$
   Solutions are 6 and 4

4. Check: if $x = 6$, $4(6)^2 - 40(6) + 96 = 0$
                        $4(36) - 240 + 96 = 0$
                        $144 - 240 + 96 = 0$
            if $x = 4$, $4(4)^2 - 40(4) + 96 = 0$
                        $64 - 160 + 96 = 0$

Example 7. Solve $-x^2 - 7x - 12 = 0$

1. Make squared term positive by multiplying both sides by -1.
   $-1(-x^2 - 7x - 12) = 0(-1)$
   $x^2 + 7x + 12 = 0$

2. Factor non zero side.

$$(x + 3)(x + 4) = 0$$

3. Set each factor to zero and solve
   $x + 3 = 0$ or $x + 4 = 0$
   $x = -3$ or $x = -4$
   Solutions are $x = -3$ and $x = -4$.

4. Check will show solutions are correct.

Example 8.   Solve $x^2 = 36$

1. Make right side equal to zero by subtracting 36 from both sides:

   $$x^2 - 36 = 0$$

2. Factor non-zero side as a difference of squares using $a^2 - b^2 = (a - b)(a + b)$.
   $x^2 - 36 = x^2 - 6^2 = (x - 6)(x + 6)$
   So $(x - 6)(x + 6) = 0$

3. Set each factor to zero and solve:
   $x - 6 = 0$ or $x + 6 = 0$
   $x = 6$ or $x = -6$
   Solutions are $x = 6$ and $x = -6$

4. Check to see if solutions are correct.

Example 9.   The product of 2 consecutive even integers is 224. What are the two integers?

Let n = first even integer. Then, $n + 2$ is the next consecutive even integer.

We have $n(n + 2) = 224$
           $n^2 + 2n = 224$

This equation is quadratic so we will set the right hand side of the equation to zero: $n^2 + 2n - 224 = 0$

Factor the non-zero side: $(n + 16)(n - 14) = 0$

Set each factor to zero and solve
$n + 16 = 0$ or $n - 14 = 0$
$n = -16$ or $n = 14$

If $n = -16$ then $n + 2 = -14$
If $n = 14$ then $n + 2 = 16$
  Thus our solutions are -16, -14 <u>or</u> 14, 16

Example 10.   The distance, d, that an object falls from rest is given by $d = 16t^2$ where d is measured in feet and t is in seconds. Find the time required for a rock to drop from a cliff which is 256 feet high.

If $d = 16t^2$ and $d = 256$ ft then $256 = 16t^2$

1. Subtract 256 from both sides to obtain a zero on the left side. (Remember, we want the coefficient of the squared term to be positive.)
$0 = 16t^2 - 256$

2. Factor out the GCF of 16.
$0 = 16(t^2 - 16)$

3. Factor $t^2 - 16$ as a difference of 2 squares.
$t^2 - 16 = (t - 4)(t + 4)$
Thus $0 = 16(t - 4)(t + 4)$

4. Set each factor **containing a variable** equal to zero and solve.
$t - 4 = 0$ or $t = 4$
$t + 4 = 0$ or $t = -4$ (not possible)
Thus, it takes 4 seconds for a rock to drop from a height of 256 feet.

Exercise Set 5.6

Solve each equation.

1. $3y^2 = 8y$
2. $a(a - 2) = 48$
3. $4m^2 + 12m = 0$
4. $6x^2 = -x + 1$
5. $4x^2 - 7x = 2$
6. $p(p - 8) = -16$
7. $6x^2 = 7x + 5$
8. $3y^2 = 5y$
9. $y^2 - 6y + 5 = 0$
10. $4w(3w + 4) = 3$
11. $t^2 - 81 = 0$
12. $x^2 = 49$
13. $3x^2 + 6x = 0$
14. $2x^2 + 5x = 12$
15. $2x^2 + x - 1 = 0$
16. $y^2 + 5y - 6 = 0$
17. $5x^2 - x = 0$
18. $6x^2 = 24$
19. $3x^2 + 14x - 5 = 0$
20. $2x^2 - 50 = 0$
21. A right triangle has a hypotenuse of 20 ft. One of the legs of

the triangle is 4 feet shorter than the other leg. Find the length of the shorter leg.

22. The product of 2 consecutive integers is 195. What are the integers?

23. The length of a rectangle is 6 less than three times the width. If the area of the rectangle is 144 sq meters, find the dimensions of the rectangle.

24. Find two numbers whose sum is 15 and whose product is 36.

Answers to exercise set 5.6

1. $y = 0, y = 8/3$
2. $a = -6, a = 8$
3. $m = 0, m = -3$
4. $x = 1/3, x = -1/2$
5. $x = 2, x = -1/4$
6. $p = 4$
7. $x = -1/2, x = 5/3$
8. $y = 0, y = 5/3$
9. $y = 1, y = 5$
10. $w = 1/6, w = -3/2$
11. $t = 9, t = -9$
12. $x = 7, x = -7$
13. $x = -2, x = 0$
14. $x = -4, x = 3/2$
15. $x = -1, x = 1/2$
16. $y = -6, y = 1$
17. $x = 0, x = 1/5$
18. $x = 2, x = -2$
19. $x = -5, x = 1/3$
20. $x = 5, x = -5$
21. 12 feet
22. 13, 15 or -13, -15
23. 8m by 18m
24. 3 and 12

Practice Test Chapter 5

1. Find the greatest common factor of 24 and 80.

2. Find the greatest common factor of $108x^3y^6$ and $24x^4y^2$.

3. Factor by grouping: $xy - 3x + xy - 3c$.

4. Factor completely: $ac - cd - 2a + 2d$.

5. Factor completely: $10p^2q^2 - 25pq^3 + 5pq^2$.

6. Factor completely: $3ay - 9y^2$.

7. Factor completely: $x^2 + 2x - 63$.

8. Factor completely: $x^2 - 14x - 32$.

9. Factor completely: $8x^2 - 10x - 7$.

10. Factor completely: $6y^3 - 36y^2 + 54y$.

11. Factor completely: $50x^3 - 2x^5$.

12. Factor completely: $125 - a^3$.

13. Factor completely: $32 - 200b^2$.

14. Factor completely: $a^3 - 8b^3$.

15. Factor completely: $6w^3 - 54w$.

16. Solve: $3x^2 + 14x = 0$

17. Solve $x^2 - x = 20$

18. Solve $3x^2 - 20x = 7$

19. Find 2 consecutive odd integers whose product is 143.

20. The perimeter of a rectanglar rug is 52 feet and its area is $168 ft^2$. Find the dimensions of the rug.

Answers to Practice Test Chapter 5

1. 8
2. $12x^3y^2$
3. $(y - 3)(x + c)$
4. $(c - 2)(a - d)$
5. $(5pq^2)(2p - 5q + 1)$
6. $3y(a - 3y)$
7. $(x - 7)(x + 9)$
8. $(x - 16)(x + 2)$
9. $(4x - 7)(2x + 1)$
10. $6y(y - 3)^2$
11. $2x^3(5 - x)(5 + x)$
12. $(5 - a)(25 + 5a + a^2)$
13. $8(2 + 5b)(2 - 5b)$
14. $(a - 2b)(a^2 + 2ab + 4b^2)$
15. $6w(w + 3)(w - 3)$
16. $\frac{-14}{3}, 0$
17. -4, 5
18. -7, 1/3
19. -13, -11 or 13, 11
20. 12' x 14'

Section 6.1   Reducing Rational Expressions

Summary:

> A **rational expression** (also called an **algebraic fraction**) is an algebraic expression of the form p/q where p and q are polynomial and q ≠ 0.
>
> For negative algebraic fraction, $\frac{-a}{b} = \frac{-a}{b} = \frac{a}{-b}$, though we generally do not write a fraction with a negative denominator.
>
> Whenever we have a rational expression containing a variable in the denominator, we always assume that the value or values of the variable that make the denominator zero are excluded.

Example 1. Determine the value or values for which the rational expression is defined:

a) $\dfrac{6x}{10x - 15}$

$$10x - 15 = 0$$
$$10x = 15$$
$$x = \frac{15}{10} = \frac{3}{2}$$

So, for all real numbers not equal to 3/2 (x ≠ 3/2).

b) $\dfrac{x^2 + 1}{x^2 - 2x - 8}$

$$x^2 - 2x - 8 = 0$$
$$(x - 4)(x + 2) = 0$$
$$x - 4 = 0 \quad \text{or} \quad x + 2 = 0$$
$$x = 4 \qquad\qquad x = -2$$

177

So, all values except 4 and -2 ($x \neq 4$ and $x \neq -2$).

Summary:

> **To Reduce Rational Expressions**
>
> 1. Factor the numerator and denominator as completely as possible.
>
> 2. Divide both numerator and denominator by any common factor.

Example 2. Reduce the following to lowest terms.

a) $\dfrac{2x^3 + 8x^2 - 6x}{2x^2}$

Factor the greatest common factor, 2x, from each term in the numerator. Since 2x is a common factor to both numerator and denominator, divide it out.

$$\frac{2x^3 + 8x^2 - 6x}{2x^2} = \frac{2x(2x^2 + 4x - 3)}{2x^2} = \frac{2x^2 + 4x - 3}{x}$$

b) $\dfrac{x^2 - 4x + 3}{x - 1}$

Factor the numerator; then divide out the common factor.

$$\frac{x^2 - 4x + 3}{x - 1} = \frac{(x - 1)(x - 3)}{x - 1} = x - 3$$

c) $\dfrac{x^2 + x - 12}{2x^2 + 7x - 4}$

Factor both the numerator and denominator, then divide out the common factors.

$$\frac{x^2 + x - 12}{2x^2 + 7x - 4} = \frac{(x - 3)(x + 4)}{(x + 4)(2x - 1)} = \frac{x - 3}{2x - 1}$$

d) $\dfrac{x^2 - 7x + 10}{5 - x}$

Factor both numerator and denominator

$$\dfrac{x^2 - 7x + 10}{5 - x} = \dfrac{(x - 5)(x - 2)}{5 - x}$$

Now, rewrite $5 - x$ as $-1(x - 5)$ so that common factors can be divide out.

$$\dfrac{(x - 5)(x - 2)}{5 - x} = \dfrac{(x - 5)(x - 2)}{-1(x - 5)} = \dfrac{x - 2}{-1} = -1(x - 2)$$

Note: A common student error is to try to divide out terms of the numerator and denominator rather than common factors.

$$\dfrac{x^2 - 3x}{x^2 - 5x} \neq \dfrac{1 - 3}{1 - 5}$$

Exercise Set 6.1

Determine the value or values of the variable when the expression is defined.

1. $\dfrac{3}{4x + 8}$ 2. $\dfrac{x - 2}{x^2 - 25}$ 3. $\dfrac{x^2}{4x^2 - 8x - 5}$

Reduce each expression to its lowest terms.

4. $\dfrac{4y - 10z}{6}$ 5. $\dfrac{x^2 + 3x - 10}{x^2 + 2x - 8}$

6. $\dfrac{x^2 + x - 12}{(x - 3)^2}$ 7. $\dfrac{x^2 - 3x - 10}{25 - x^2}$

8. $\dfrac{2x^3 + 2x^2 - 4x}{x^3 + 2x^2 - 3x}$ 9. $\dfrac{6x^2 - 7x + 2}{6x^2 + 5x - 6}$

10. $\dfrac{2n^2 - 9n + 4}{2n^2 - 5n - 12}$ 11. $\dfrac{12 - 6x}{x - 2}$

12. $\dfrac{x^2 - 36}{6 - x}$ 13. $\dfrac{x^4 + x^2}{x^2 + 1}$

14. $\dfrac{x^3 + 27}{x + 3}$ 15. $\dfrac{x^3 - 8}{x^2 + 2x + 4}$

Answers to Exercise Set 6.1

1. $x \neq -2$ 2. $x \neq 5$ and $x \neq -5$

3. $x \neq -\dfrac{1}{2}$ and $x \neq \dfrac{5}{2}$ 4. $\dfrac{2y - 5z}{3}$

5. $\dfrac{x + 5}{x + 4}$ 6. $\dfrac{x + 4}{x - 3}$

7. $\dfrac{-(x + 2)}{x + 5}$ 8. $\dfrac{2(x + 2)}{x + 3}$

9. $\dfrac{2x - 1}{2x + 3}$ 10. $\dfrac{2n - 1}{2n + 3}$

11. $x - 6$ 12. $-(x + 6)$

13. x^2 14. $x^2 - 3x + 9$

15. $x - 2$

Section 6.2 **Multiplication and Division of Rational Expression**

Summary:

> **To Multiply Rational Expressions**
>
> 1. Factor all numerators and denominators completely.
>
> 2. Divide out common factors
>
> 3. Multiply numerators together and denominators together.

Example 1. Multiply $\dfrac{8x^2}{9y^3} \cdot \dfrac{3y^2}{4x^3}$

$$\dfrac{8x^2}{9y^3} \cdot \dfrac{3y^2}{4x^3}$$

$$\dfrac{2 \cdot 2 \cdot 2 \cdot x \cdot x}{3 \cdot 3 \cdot y \cdot y \cdot y} \cdot \dfrac{3 \cdot y \cdot y}{2 \cdot 2 \cdot x \cdot x \cdot x}$$

$$\dfrac{2 \cdot 2 \cdot 2 \cdot x \cdot x}{3 \cdot 3 \cdot y \cdot y \cdot y} \cdot \dfrac{3 \cdot y \cdot y}{2 \cdot 2 \cdot x \cdot x \cdot x} \quad \text{Divide out common factors}$$

$$\dfrac{2}{3y} \cdot \dfrac{1}{x}$$

Now, multiply the remaining numerators together and the remaining denominator together.

$$\dfrac{2}{3y} \cdot \dfrac{1}{x} = \dfrac{2}{3xy}$$

Example 2. Multiply $\dfrac{12a^5b}{25x^2y^5} \cdot \dfrac{20xy^2}{16a^3b^3}$

$$\frac{3 \cdot 4 \cdot a^3 \cdot a^2 \cdot b}{5 \cdot 5 \cdot x \cdot xy^2 \cdot y^3} \cdot \frac{5 \cdot 4 \cdot x \cdot y^2}{4 \cdot 4 \cdot a^3 \cdot b \cdot b^2} = \frac{3a^2}{5xy^3 b^2}$$

Example 3. $\dfrac{8x - 12}{14x + 7} \cdot \dfrac{42x + 21}{32x - 48}$

$$\frac{8x - 12}{14x + 7} \cdot \frac{42x + 21}{32x - 48}$$

$$\frac{4(2x - 3)}{7(x + 2)} \cdot \frac{21(x + 2)}{16(2x - 3)} = \frac{3}{4}$$

Example 4. $\dfrac{x^2 + 3x}{x^2 - 3x - 4} \cdot \dfrac{x^2 - 5x + 4}{x^2 + 2x - 3}$

$$\frac{x(x + 3)}{(x - 4)(x + 1)} \cdot \frac{(x - 4)(x - 1)}{(x + 3)(x - 1)} = \frac{x}{x + 1}$$

Example 5. $\dfrac{x^2 + x - 6}{x^2 + 7x + 12} \cdot \dfrac{x^2 + 3x - 4}{4 - x^2}$

$$\frac{(x + 3)(x - 2)}{(x - 3)(x + 4)} \cdot \frac{(x + 4)(x - 1)}{(2 - x)(2 + x)}$$

$\dfrac{(x - 2)(x - 1)}{(2 - x)(2 + x)}$ Recall: $2 - x$ can be written as
$-1(x - 2)$

$\dfrac{(x - 2)(x - 1)}{-1(x - 2)(2 + x)} = \dfrac{-(x - 1)}{2 + x}$ or $-\dfrac{x - 1}{x + 2}$

Summary:

> **Division**
>
> $a/b \div c/d = a/b \cdot d/c = ad/bc$, $b \neq 0$, $d \neq 0$, $c \neq 0$.
>
> **To Divide Rational Expressions**
>
> Invert the divisor (the second fraction) and multiply.

Example 6. Divide $\dfrac{2x^2y^3}{5a^3b} \div \dfrac{8xy^5}{25ab^4}$

$$\dfrac{2x^2y^3}{5a^3b} \div \dfrac{8xy^5}{25ab^4} = \dfrac{2x^2y^3}{5a^3b} \cdot \dfrac{25ab^4}{8xy^5}$$

$$= \dfrac{2 \cdot x \cdot x \cdot y^3}{5a^2 \cdot a \cdot b} \cdot \dfrac{5 \cdot 5 \cdot a \cdot b^3 \cdot b}{2 \cdot 4 \cdot x \cdot y^3 \cdot y^2} = \dfrac{10xb^3}{4a^2y^2}$$

Example 7. Divide $\dfrac{5a^2y + 3a^2}{2x^3 + 5x^2} \div \dfrac{10ay + 6a}{6x^3 + 15x^2}$

$$\dfrac{5a^2y + 3a^2}{2x^3 + 5x^2} \div \dfrac{10ay + 6a}{6x^3 + 15x^2} = \dfrac{5a^2y + 3a^2}{2x^3 + 5x^2} \cdot \dfrac{6x^3 + 15x^2}{10ay + 6a}$$

$$\dfrac{a^2(5y + 3)}{x^2(2x + 5)} \cdot \dfrac{3x^2(2x + 5)}{2a(5y + 3)} = \dfrac{3a}{2}$$

Example 8. $\dfrac{x^2 - x - 2}{x^2 - 7x + 10} \div \dfrac{x^2 - 3x - 4}{40 - 3x - x^2}$

$$\dfrac{x^2 - x - 2}{x^2 - 7x + 10} \div \dfrac{x^2 - 3x - 4}{40 - 3x - x^2} = \dfrac{x^2 - x - 2}{x^2 - 7x + 10} \cdot \dfrac{40 - 3x - x^2}{x^2 - 3x - 4}$$

$$\frac{(x-2)(x+1)}{(x-5)(x-2)} \cdot \frac{(8+x)(5-x)}{(x-4)(x+1)} = \frac{-1(8+x)}{x-4}$$

Exercise Set 6.2

Multiply

1. $\dfrac{m^3}{2} \cdot \dfrac{4m}{m^4}$

2. $\dfrac{6y^5x^6}{y^3x^4} \cdot \dfrac{y^4x^2}{3y^5x^7}$

3. $\dfrac{6p^2q^3}{4p^3q} \cdot \dfrac{18p^2q^3}{12p^3q^3}$

4. $\dfrac{2r}{8r+4} \cdot \dfrac{14r+7}{3}$

5. $\dfrac{6a-10}{5a} \cdot \dfrac{3}{9a-15}$

6. $\dfrac{m^2+6m+9}{m} \cdot \dfrac{m^2}{m^2-9}$

7. $\dfrac{x^2-1}{2x} \cdot \dfrac{1}{1-x}$

8. $\dfrac{6r-5s}{3r+2s} \cdot \dfrac{6r+4s}{5s-6r}$

9. $\dfrac{2r^2+5r-3}{r^2-1} \cdot \dfrac{r^2-2r+1}{2r^2-7r+3}$

10. $\dfrac{a^2-16}{a^2+a-12} \cdot \dfrac{a^2+5a+6}{a^2-2a-8}$

Divide

11. $\dfrac{6a^4}{a^3} \div \dfrac{12a^2}{a^5}$

12. $\dfrac{25a^2b}{60a^3b^2} \div \dfrac{5a^4b^2}{16a^2b}$

13. $\dfrac{8s^4t^2}{5t^6} \div \dfrac{s^3t^2}{10s^2t^4}$

14. $\dfrac{7k+7}{2} \cdot \dfrac{4k+4}{5}$

15. $\dfrac{p^2-36}{p+1} \div \dfrac{6-p}{p}$

16. $\dfrac{x^2+3x-40}{x^2+2x-35} \div \dfrac{x^2+2x-48}{x^2+3x-18}$

17. $\dfrac{y^2 - y - 56}{y^2 + 8y + 7} \div \dfrac{y^2 - 13y + 40}{y^2 - 4y - 5}$

18. $\dfrac{8 + 2x - x^2}{x^2 + 7x + 10} \div \dfrac{x^2 - 11x + 28}{x^2 - x - 42}$

19. $\dfrac{2x^2 - 3x - 20}{2x^2 - 7x - 30} \div \dfrac{2x^2 - 5x - 12}{4x^2 + 12x + 9}$

20. $\dfrac{9x^2 - 16}{6x^2 - 11x + 4} \div \dfrac{6x^2 + 11x + 4}{8x^2 + 10x + 3}$

Answers to Exercise Set 6.2

1. 2
2. $\dfrac{2y}{x^3}$
3. $\dfrac{9q^3}{4p^2}$
4. $\dfrac{7r}{6}$
5. $\dfrac{6}{5a}$
6. $\dfrac{m(m + 3)}{m - 3}$
7. $\dfrac{-(x + 1)}{2x}$
8. -2
9. $\dfrac{(r + 3)(r - 1)}{(r + 1)(r - 3)}$
10. $\dfrac{a + 3}{a - 3}$
11. $\dfrac{a^4}{2}$
12. $\dfrac{4}{3a^3b^2}$
13. $\dfrac{16s^3}{t^2}$
14. $\dfrac{35}{8}$
15. $\dfrac{-p(p + 6)}{p + 1}$
16. $\dfrac{(x + 6)(x - 3)}{(x + 7)(x - 6)}$
17. 1
18. $\dfrac{-(x + 6)}{x + 5}$
19. $\dfrac{2x + 3}{x - 6}$
20. $\dfrac{4x + 3}{2x - 1}$

Section 6.3 **Addition and Subtraction of Rational Expressions with a Common Denominator and Finding the Least Common Denominator**

Summary:

Addition and Subtraction

$\dfrac{a}{c} + \dfrac{b}{c} = \dfrac{a+b}{c}, \; c \neq 0$

$\dfrac{a}{c} - \dfrac{b}{c} = \dfrac{a-b}{c}, \; c \neq 0$

Summary:

To Add or Subtract Rational Expressions with a Common Denominator

1. Add or subtract the numerator

2. Place the sum or differences of the numerator found in step 1 over the common denominator

3. Reduce the fraction if possible

Example 1. Add $\dfrac{4}{3t} + \dfrac{9}{3t}$

$$\dfrac{4}{3t} + \dfrac{9}{3t} = \dfrac{13}{3t}$$ (Added numerator) Cannot be reduced

Example 2. Add $\dfrac{x^2}{x+2} + \dfrac{2x}{x+2}$

$$\dfrac{x^2}{x+2} + \dfrac{2x}{x+2} = \dfrac{x^2 + 2x}{x+2} \quad \text{(Added numerator)}$$

$$= \dfrac{x(x+2)}{x+2} = x \quad \text{(Reduced fraction)}$$

Example 3. Add $\dfrac{a^2 + a - 5}{(a-1)(a+1)} + \dfrac{4-a}{(a-1)(a+1)}$

$$\dfrac{a^2 + a - 5}{(a-1)(a+1)} + \dfrac{4-a}{(a-1)(a+1)} = \dfrac{a^2 + a - 5 + (4-a)}{(a-1)(a+1)}$$
(Added numerator)

$$= \dfrac{a^2 - 1}{(a-1)(a+1)}$$

$$= \dfrac{(a-1)(a+1)}{(a-1)(a+1)} \quad \text{(Reduced fractions)}$$

Example 4. Subtract $\dfrac{3y}{y-2} - \dfrac{6}{y-2}$

$$\dfrac{3y}{y-2} - \dfrac{6}{y-2} = \dfrac{3y - 6}{y-2} \quad \text{(Subtracted numerator)}$$

$$= \dfrac{3(y-2)}{(y-2)} = 3 \quad \text{(Reduced fractions)}$$

Example 5. Subtract $\dfrac{a^2}{a^2 - 2a - 3} - \dfrac{6a - 9}{a^2 - 2a - 3}$

$$\dfrac{a^2}{a^2 - 2a - 3} - \dfrac{6a - 9}{a^2 - 2a - 3} = \dfrac{a^2 - (6a - 9)}{a^2 - 2a - 3}$$

$$= \dfrac{a^2 - 6a + 9}{a^2 - 2a - 3} \quad \text{(Subtracted numerators)}$$

$$= \dfrac{(a - 3)(a - 3)}{(a - 3)(a + 1)}$$

$$= \dfrac{a - 3}{a + 1} \quad \text{(Reduced fractions)}$$

Note: A common student error is to fail to change the signs of all the terms of the numerator being subtracted.

Summary:

> **To Find the Least Common Denominator of Rational Expressions**
>
> 1. Factor each denominator completely. Any factor that occurs more than once should be expressed as powers. For example, $(x + 5)(x + 5)$ should be expressed as $(x + 5)^2$.
>
> 2. List all different factors (other than 1) that appear in any of the denominators. When the factor appears in more than one denominator, write that factor with the highest power that appears.
>
> 3. The least common denominator is the product of all the factors determined in the second step.

Example 6. Find the least common denominator of the following:

a) $\dfrac{2}{3x} + \dfrac{5}{2}$

   The common factors are 3, x, and 2. List each factor with its highest power. The least common denominator (LCD) is the product of these factors.

   LCD = 3 · 2 · x = 6x

b) $\dfrac{4a + b}{4a^2b^3} + \dfrac{2a - b^2}{ab^4}$

   The common factor (other than 1) are 4, a, and b. List each factor with its highest power. The LCD is the product of these factors.

   LCD = 4 · a^2 · b^4 = $4a^2b^4$

c) $\dfrac{t}{4t + 8} - \dfrac{t^2}{3t + 6}$

   Here, factor both denominator

   $\dfrac{t}{4(t + 2)} - \dfrac{t^2}{3(t + 2)}$

   Common factors (other than 1) are 4, 3, and (t + 2).
   LCD = 4 · 3 · (t + 2) = 12(t + 2)

d) $\dfrac{2x}{x^2 + 6x + 9} + \dfrac{4}{x^2 - 2x - 15}$

   Here, factor both denominators

   $\dfrac{2x}{(x + 3)^2} + \dfrac{4}{(x + 3)(x - 5)}$

   Common factors (other than 1) are (x + 3) and (x - 5)
   List each factor with highest power
   LCD = $(x + 3)^2$ and (x - 5)

Exercise Set 6.3

Add or Subtract.

1. $\dfrac{4}{y-3} + \dfrac{6y}{y-3}$

2. $\dfrac{5a+1}{2a+3} + \dfrac{5a+14}{2a+3}$

3. $\dfrac{5x}{18} + \dfrac{7x}{18}$

4. $\dfrac{x}{x^2-1} + \dfrac{1}{x^2-1}$

5. $\dfrac{2x}{x-2} - \dfrac{4}{x-2}$

6. $\dfrac{3x-1}{x^2-5x+4} - \dfrac{2x+3}{x^2-5x+4}$

7. $\dfrac{2n}{3n+4} - \dfrac{5n-3}{3n+4}$

8. $\dfrac{3x-1}{x^2+5x-6} - \dfrac{2x-7}{x^2+5x-6}$

9. $\dfrac{4y+7}{2y^2+7y-4} - \dfrac{y-5}{2y^2+7y-4}$

10. $\dfrac{x+1}{2x^2-5x-12} + \dfrac{x+2}{2x^2-5x-12}$

Find the least common denominator

11. $\dfrac{x}{2} - \dfrac{y}{x}$

12. $\dfrac{6}{5y} + \dfrac{7}{9y}$

13. $\dfrac{2}{y^2} - \dfrac{5}{y^3}$

14. $\dfrac{5}{t^2} + \dfrac{2}{2t}$

15. $\dfrac{10}{6b^3} + \dfrac{2}{8b^2}$

16. $\dfrac{4}{15a^2b^2} - \dfrac{12}{5ab^4}$

17. $\dfrac{x+1}{4x^2} + \dfrac{x-3}{6x^2-12x}$

18. $\dfrac{2x-1}{2x-x^2} - \dfrac{x}{x^2-x-6}$

19. $\dfrac{3}{x^2+x-2} + \dfrac{x}{x+2}$

20. $\dfrac{x}{x^2-16} - \dfrac{4}{x^2+8x+16}$

Answers to Exercise Set 6.3

1. $\dfrac{6y + 4}{y - 3}$ 2. 3 3. $\dfrac{2x}{3}$

4. $\dfrac{1}{x - 1}$ 5. 2 6. $\dfrac{1}{x - 1}$

7. $\dfrac{-3n + 3}{3n + 4}$ 8. $\dfrac{1}{x - 1}$ 9. $\dfrac{3}{2y - 1}$

10. $\dfrac{1}{x - 4}$ 11. $2x$ 12. $45y$

13. y^3 14. $2t^2$ 15. $24b^3$

16. $15a^2b^4$ 17. $12x^2(x - 2)$ 18. $x(x - 2)(x + 3)$

19. $(x + 2)(x - 1)$ 20. $(x + 4)^2(x - 4)$

Set 6.4   Addition and Subtraction of Rational Expressions

Summary:

> **To Add or Subtract Two Rational Expressions with Unlike Denominators**
>
> 1. Determine the LCD
>
> 2. Rewrite each fraction as an equivalent fraction with the LCD. This is done by multiplying both the numerator and denominator of each fraction by any factors needed to obtain the LCD.
>
> 3. Add or subtract the numerator while maintaining the LCD.
>
> 4. When possible, factor the remaining numerator and reduce the fraction.

Example 1.   Add $\dfrac{5}{x} + \dfrac{3}{x^2}$

   (Step 1)   LCD = x^2

   (Step 2) $\dfrac{5 \cdot x}{x \cdot x} + \dfrac{3}{x^2} = \dfrac{5x}{x^2} + \dfrac{3}{x^2}$

   (Step 3) $\dfrac{5x}{x^2} + \dfrac{3}{x^2} = \dfrac{5x + 3}{x^2}$

Example 2.   Subtract $\dfrac{3}{4xy} - \dfrac{1}{2x}$

   (Step 1)   LCD = $4xy$

(Step 2) $\quad \dfrac{3}{4xy} - \dfrac{1 \cdot (2y)}{(2x)(2y)} = \dfrac{3}{4xy} - \dfrac{2y}{4xy}$

(Step 3) $\quad \dfrac{3}{4xy} - \dfrac{2y}{4xy} = \dfrac{3 - 2y}{4xy}$

Example 3. Add $\dfrac{2x}{x-3} + \dfrac{5}{x+5}$

(Step 1) LCD = $(x-3)(x+5)$

(Step 2) $\quad \dfrac{2x(x+5)}{(x-3)(x+5)} + \dfrac{5(x-3)}{(x+5)(x-3)}$

(Step 3) $\quad \dfrac{2x^2 + 10x + 5x - 15}{(x-3)(x+5)} = \dfrac{2x^2 + 15x - 15}{(x-3)(x+5)}$

Example 4. Add $\dfrac{x^2}{x-3} + \dfrac{9}{3-x}$

(Step 1) $x - 3$ since $3 - x = -1(x-3)$

(Step 2) $\quad \dfrac{x^2}{x-3} + \dfrac{(-1)(9)}{(-1)(3-x)} = \dfrac{x^2}{x-3} + \dfrac{9}{x-3}$

(Step 3) $\quad \dfrac{x^2}{x-3} + -\dfrac{9}{x-3} = \dfrac{x^2 - 9}{x-3}$

(Step 4) $\quad \dfrac{x^2 - 9}{x-3} = \dfrac{(x-3)(x+3)}{(x-3)} = x + 3$

Example 5. Subtract $\dfrac{x}{x^2 - 2x - 3} - \dfrac{5}{x^2 - 1}$

(Step 1) LCD = $(x - 3)(x - 1)(x + 1)$

Here, $x^2 - 2x - 3 = (x - 3)(x + 1)$

$x^2 - 1 = (x - 1)(x + 1)$

(Step 2) $\dfrac{x}{(x - 3)(x + 1)} - \dfrac{5}{(x - 1)(x + 1)} =$

$= \dfrac{x(x - 1)}{(x - 3)(x + 1)(x - 1)} - \dfrac{5(x - 3)}{(x - 1)(x + 1)(x - 3)}$

(Step 3) $\dfrac{x^2 - x}{(x - 3)(x + 1)(x - 1)} + \dfrac{-5x + 15}{(x - 1)(x + 1)(x - 3)}$

$= \dfrac{x^2 - 6x + 15}{(x - 3)(x + 1)(x - 1)}$

Exercise Set 6.4

Add or Subtract

1) $\dfrac{x}{8} - \dfrac{y}{12}$

2) $\dfrac{x + 3}{10} + \dfrac{3x - 1}{15}$

3) $\dfrac{7}{a} + \dfrac{5}{b}$

4) $\dfrac{4x - 3}{6x} + \dfrac{2x + 3}{4x}$

5) $\dfrac{2x - 3}{2x} + \dfrac{x + 3}{3x}$

6) $\dfrac{3y - 2}{12y} - \dfrac{y - 3}{18y}$

7) $4 + \dfrac{5a}{a + 3}$

8) $\dfrac{x + 3}{6x} - \dfrac{x - 3}{8x^2}$

9) $\dfrac{x + 5}{3x^2} + \dfrac{2x + 1}{2x}$

10) $\dfrac{2}{x - 3} + \dfrac{5}{x - 4}$

11) $\dfrac{3}{y+6} - \dfrac{4}{y-3}$

12) $\dfrac{3x}{x-4} + \dfrac{2}{x+6}$

13) $\dfrac{2b}{b-7} + \dfrac{5}{7-b}$

14) $\dfrac{4y}{6-y} + \dfrac{5}{y-6}$

15) $\dfrac{x}{x^2-9} - \dfrac{3}{x-3}$

16) $\dfrac{2x}{x^2-x-6} - \dfrac{3}{x+2}$

17) $\dfrac{x^2-2}{2x^2-x-3} - \dfrac{x-2}{2x-3}$

18) $\dfrac{x+2}{2x^2+5x+2} - \dfrac{3}{2x+1}$

Answers to Exercise Set 6.4

1) $\dfrac{3x-2y}{24}$

2) $\dfrac{9x+7}{30}$

3) $\dfrac{7b+5a}{ab}$

4) $\dfrac{14x+3}{12x}$

5) $\dfrac{8x-3}{6x}$

6) $\dfrac{7}{36}$

7) $\dfrac{3(3a+4)}{a+3}$

8) $\dfrac{4x^2+9x+9}{24x^2}$

9) $\dfrac{6x^2+5x+10}{6x^2}$

10) $\dfrac{7x-23}{(x-3)(x-4)}$

11) $\dfrac{-y-33}{(y+6)(y-3)}$

12) $\dfrac{3x^2+20x+8}{(x-4)(x+6)}$

13) $\dfrac{2b-5}{b-7}$

14) $\dfrac{-4y+5}{y-6}$

15) $\dfrac{4x+9}{(x+3)(x-3)}$

16) $\dfrac{-x+9}{(x-3)(x+2)}$

17) $\dfrac{x}{(2x-3)(x+1)}$

18) $\dfrac{-2}{2x+1}$

Set 6.5   Complex Fractions

Summary:

> A **complex fraction** is one that has a fraction in its numerator or its denominator or in both its numerator and denominator.
>
> $\dfrac{a+b}{a}$ numerator of complex fraction
>
> ——————   <————————> main fraction line
>
> $\dfrac{a-b}{b}$ denominator of complex fraction

Summary:

> **To Simplify a Complex Fraction by Combining Terms**
>
> 1. Add or Subtract the fractions in both the numerator and denominator of the complex fraction to obtain single fractions in both the numerator and the denominator.
>
> 2. Invert the denominator of the complex fraction and multiply the numerator by it.
>
> 3. Simplify further if possible.

Example 1. Simplify $\dfrac{\dfrac{2}{3} - \dfrac{1}{8}}{\dfrac{3}{4} + \dfrac{1}{3}}$

Solution.

(Step 1) $\quad \dfrac{\dfrac{2}{3} \cdot \dfrac{8}{8} - \dfrac{1}{8} \cdot \dfrac{3}{3}}{\dfrac{3}{4} \cdot \dfrac{3}{3} + \dfrac{1}{3} \cdot \dfrac{4}{4}} \quad \begin{array}{l} LCD = 8 \cdot 3 = 24 \\[6pt] LCD = 4 \cdot 3 = 12 \end{array}$

$\dfrac{\dfrac{16}{24} - \dfrac{3}{24}}{\dfrac{9}{12} + \dfrac{4}{12}} = \dfrac{\dfrac{13}{24}}{\dfrac{13}{12}}$

(Step 2) $\quad \dfrac{13}{24} \cdot \dfrac{12}{13} = \dfrac{13}{24} \cdot \dfrac{12}{13} = \dfrac{1}{2}$

Example 2. Simplify $\dfrac{1 - \dfrac{4}{x^2}}{1 - \dfrac{2}{x}}$

Solution.

(Step 1) $\quad \dfrac{\dfrac{1}{1} \cdot \dfrac{x^2}{x^2} - \dfrac{4}{x^2}}{\dfrac{1}{1} \cdot \dfrac{x}{x} - \dfrac{2}{x}} \quad \begin{array}{l} LCD = x^2 \\[6pt] LCD = x \end{array}$

$\dfrac{\dfrac{x^2}{x^2} - \dfrac{4}{x^2}}{\dfrac{x}{x} - \dfrac{2}{x}} = \dfrac{\dfrac{x^2 - 4}{x^2}}{\dfrac{x - 2}{x}}$

(Step 2) $\quad \dfrac{x^2 - 4}{x^2} \cdot \dfrac{x}{x - 2} = \dfrac{(x - 2)(x + 2)}{x^2} \cdot \dfrac{x}{(x - 2)} = \dfrac{x + 2}{x}$

Summary:

> **To Simplify a Complex Fraction Using Multiplication First**
>
> 1. Find the least common denominator of all the denominator appearing in the complex fraction.
>
> 2. Multiply both the numerator and denominator of the complex fraction by the LCD found in step 1.
>
> 3. Simplify when possible.

Example 3.   Simplify $\dfrac{\frac{2}{3} - \frac{1}{8}}{\frac{3}{4} + \frac{1}{3}}$

(Step 1)   LCD of all denominators = $2^3 \cdot 3 = 24$

(Step 2) $\dfrac{24}{24} \cdot \dfrac{\left(\frac{2}{3} - \frac{1}{8}\right)}{\left(\frac{3}{4} + \frac{1}{3}\right)}$

$$\dfrac{(24)\left(\frac{2}{3}\right) - (24)\left(\frac{1}{8}\right)}{(24)\left(\frac{3}{4}\right) + (24)\left(\frac{1}{3}\right)} = \dfrac{16 - 3}{18 + 8} = \dfrac{13}{26}$$

(Step 3) $\dfrac{13}{26} = \dfrac{13 - 1}{13 - 2} = \dfrac{1}{2}$

Example 4.   Simplify $\dfrac{1 - \dfrac{4}{x^2}}{11 - \dfrac{2}{x}}$

(Step 1)   LCD of all denominators = x^2

(Step 2) $\dfrac{x^2 \left(1 - \dfrac{4}{x^2}\right)}{x^2 \left(1 - \dfrac{2}{x}\right)}$

$$\dfrac{(x^2)(1) - x^2\left(\dfrac{4}{x^2}\right)}{x^2 - x^2\left(\dfrac{2}{x}\right)} = \dfrac{x^2 - 4}{x^2 - 2x}$$

(Step 3) $\dfrac{x^2 - 4}{x^2 - 2x} = \dfrac{(x - 2)(x + 2)}{x(x - 2)} = \dfrac{x + 2}{x}$

Exercise Set 6.5

Simplify each expression

1. $\dfrac{3 - \dfrac{2}{3}}{3 + \dfrac{1}{6}}$

2. $\dfrac{\dfrac{1}{5} - \dfrac{1}{2}}{\dfrac{2}{5} + \dfrac{3}{10}}$

3. $\dfrac{5 - \dfrac{2}{4}}{3 + \dfrac{1}{3}}$

4. $\dfrac{\dfrac{3a^2 b}{2c^3}}{\dfrac{4ab^4}{6c^4}}$

5. $\dfrac{\dfrac{2t^2u^4}{5r^4}}{\dfrac{8tu^2}{15r}}$

6. $\dfrac{1 + \dfrac{3}{x}}{1 - \dfrac{4}{y}}$

7. $\dfrac{\dfrac{1}{x} + 2}{\dfrac{1}{x} + 3}$

8. $\dfrac{\dfrac{1}{a} - \dfrac{2}{b}}{\dfrac{3}{a} - \dfrac{5}{b}}$

9. $\dfrac{\dfrac{x+1}{x-1}}{\dfrac{x+1}{x}}$

10. $\dfrac{\dfrac{b-1}{b}}{\dfrac{b^2-1}{4}}$

11. $\dfrac{\dfrac{x}{y} - \dfrac{y}{x}}{\dfrac{x}{y} + \dfrac{y}{x}}$

12. $\dfrac{\dfrac{x}{y} - \dfrac{y}{x}}{\dfrac{y}{x} - \dfrac{x}{y}}$

Answers to Exercise Set 6.5

1. $\dfrac{14}{19}$

2. $-\dfrac{3}{7}$

3. $\dfrac{43}{30}$

4. $\dfrac{9ac}{4b^3}$

5. $\dfrac{3tu^3}{4r^3}$

6. $\dfrac{xy + 3y}{xy - 4x}$

7. $\dfrac{1 + 2x}{1 + 3x}$

8. $\dfrac{b - 2a}{3b - 5a}$

9. $\dfrac{x}{x - 1}$

10. $\dfrac{4}{b(b + 1)}$

11. $\dfrac{x^2 - y^2}{x^2 + y^2}$

12. -1

Section 6.6   Solving Rational Expression

Summary:

> **To Solve Rational Equations**
>
> 1. Determine the LCD of all fractions in the equation.
>
> 2. Multiply **both** sides of the equation by the LCD. **This will result in every term in the equation being multiplied by the LCD.**
>
> 3. Remove any parentheses and combine like terms on each side of the equation.
>
> 4. Solve the equation using the properties discussed in earlier chapters.
>
> 5. Check your solution in the original equation.

Example 1   Solve $\frac{x}{2} - \frac{x}{3} = 10$

Solution.
(step 1)   LCD = 2 · 3 = 6

(Step 2) $6\left(\frac{x}{2} - \frac{x}{3}\right) = 6(10)$

(Step 3) $\left(6 \cdot \frac{x}{2}\right) - \left(6 \cdot \frac{x}{3}\right) = 6(10)$

(Step 4)   3x − 2x = 60
            x = 60

(Step 5)   3x − 2x   = 60
            3(60) − 2(60) = 60
            180 − 120   = 60
            60   = 60 True   60 is solution

Example 2. $$\frac{3x + 1}{4} = \frac{13 - x}{2}$$

Solution.
(Step 1)   LCD = 4

(Step 2) $$4\left(\frac{3x + 1}{4}\right) = 4\left(\frac{13 - x}{2}\right)$$

(Step 3) $$4 \cdot \frac{3x + 1}{4} = 4 \cdot \frac{13 - x}{2}$$
$$3x + 1 = 2(13 - x)$$
$$3x + 1 = 26 - 2x$$

(Step 4) $$3x + 1 = 26 - 2x$$
$$5x + 1 = 26$$
$$5x = 25$$
$$x = 5$$

(Step 5) $$\frac{3x + 1}{4} = \frac{13 - x}{2}$$

$$\frac{3(5) + 1}{4} = \frac{13 - 5}{2}$$

$$\frac{15 + 1}{4} = \frac{8}{2}$$

$$\frac{16}{4} = \frac{8}{2}$$

4 = 4 True   5 is solution

Example 3. Solve $\frac{x + 9}{2} = \frac{15}{x} - 3$

Solution.
(Step 1)   LCD = 2x

(Step 2) $$2x\left(\frac{x + 9}{2x}\right) = 2x\left(\frac{15}{x} - 3\right)$$

(Step 3) $2x \cdot \dfrac{x + 9}{2x} = 2x \cdot \dfrac{15}{x} - 2x(3)$

$x + 9 = 30 - 6x$

(Step 4) $7x + 9 = 30$
$7x = 21$
$x = 3$

(Step 5) $\dfrac{x + 9}{2x} = \dfrac{15}{x} - 3$

$\dfrac{3 + 9}{2(3)} = \dfrac{15}{3} - 3$

$\dfrac{12}{6} = 5 - 3$

$2 = 2$ 2 is a solution

Example 4. $\dfrac{x}{x + 2} + \dfrac{2}{x - 2} = \dfrac{x + 6}{x^2 - 4}$

Solution.
(Step 1)   LCD = $(x + 2)(x - 2)$

(Step 2) $(x + 2)(x - 2)\left(\dfrac{x}{x + 2} + \dfrac{2}{x - 2}\right) =$

$(x + 2)(x - 2)\left(\dfrac{x + 6}{x^2 - 4}\right)$

(Step 3) $(x + 2)(x - 2)\left(\dfrac{x}{x + 2}\right) + (x + 2)(x - 2)\left(\dfrac{2}{x - 2}\right) =$

$(x + 2)(x - 2)\left(\dfrac{x + 6}{x^2 - 4}\right)$

$x(x - 2) + 2(x + 2) = x + 6$
$x^2 - 2x + 2x + 4 = x + 6$
$x^2 + 4 = x + 6$

(Step 4) $\quad x^2 - x - 2 = 0$
$\quad\quad\quad\quad (x - 2)(x + 1) = 0$

$\quad\quad\quad\quad x - 2 = 0 \quad$ or $\quad x + 1 = 0$
$\quad\quad\quad\quad x = 2 \quad\quad\quad\quad\quad x = -1$

(Step 5) $\quad$ Check $x = -1$

$$\frac{x}{x+2} + \frac{2}{x-2} = \frac{x+6}{x^2-4}$$

$$\frac{-1}{-1+2} + \frac{2}{-1-2} = \frac{-1+6}{(-1)^2 - 4}$$

$$-\frac{1}{1} + \frac{2}{-3} = \frac{5}{1-4}$$

$$-1 + -\frac{2}{3} = -\frac{5}{3}$$

$$-\frac{5}{3} = -\frac{5}{3} \quad\quad \text{So, } -1 \text{ is a solution}$$

Check $x = 2$

$$\frac{x}{x+2} + \frac{2}{x-2} = \frac{x+6}{x^2-4}$$

$$\frac{2}{2+2} + \frac{2}{2-2} = \frac{2+6}{2^2-4}$$

$$\frac{2}{4} + \frac{2}{0} = \frac{8}{0} \quad (2/0, 8/0 \text{ are not real number})$$

Therefore, 2 is extraneous.

Exercise set 6.6

Solve each equation and check your solution

1. $\quad \dfrac{6}{2a+1} = 2$
2. $\quad \dfrac{9}{2x-5} = -2$

3. $\quad \dfrac{2x}{3} - \dfrac{5}{2} = -\dfrac{1}{2}$
4. $\quad \dfrac{x}{3} - \dfrac{1}{4} = \dfrac{1}{12}$

5. $\dfrac{2x-5}{8} + \dfrac{1}{4} = \dfrac{x}{8} + \dfrac{3}{4}$ 6. $\dfrac{3x+4}{12} - \dfrac{1}{3} = \dfrac{5x+2}{12} - \dfrac{1}{12}$

7. $2 + \dfrac{5}{x} = 7$ 8. $3 + \dfrac{8}{n} = 5$

9. $\dfrac{3}{x-2} = \dfrac{4}{x}$ 10. $\dfrac{5}{x+3} = \dfrac{3}{x-1}$

11. $2 + \dfrac{3}{a-3} = \dfrac{a}{a-3}$ 12. $\dfrac{2x}{x+4} = \dfrac{3}{x-1}$

13. $x + \dfrac{6}{x-2} = \dfrac{3x}{x-2}$ 14. $\dfrac{x+2}{x^2-x-2} + \dfrac{x+1}{x^2-4} = \dfrac{1}{x+1}$

Answers to Exercise Set 6.6

1. 1   2. $\dfrac{1}{4}$ 3. 3   4. 1

5. 9   6. 2   7. 1   8. 2

9. 8   10. 7   11. no solution

12. $-\dfrac{3}{2}$, 4   13. 3   14. −3

Set 6.7 Applications of Rational Equations

Many applications of algebra involve rational equations. After we set up the application as an equation, we solve the rational equation as we did in Section 6.6.

Example 1. One number is three times the other. If the sum of the reciprocal of the number is 12, find the number.

Solution. Let x = one number
$3x$ = other number

$$\frac{1}{x} + \frac{1}{3x} = 12$$

reciprocal   reciprocal
of one number   of the other

$$3x\left(\frac{1}{x} + \frac{1}{3}x\right) = 3x(12)$$

$$3 + 1 = 36x$$
$$4 = 36x$$

$$\frac{4}{36} = x$$

$$\frac{1}{9} = x$$

One number is $\frac{1}{9}$ and the other number is $3\left(\frac{1}{9}\right) = \frac{3}{9} = \frac{1}{3}$

Example 2. The area of a triangle is 20 square feet. If the base is 3 feet longer than the height, find the base and height.

Solution. Let x = height
$x + 3$ = base

Area = $\frac{1}{2}$ · base · height

$$20 = \frac{1}{2}(x + 3)(x)$$

$$2 \cdot 20 = 2 \cdot \frac{1}{2} \cdot (x + 3)(x)$$
$$40 = (x + 3)(x)$$
$$40 = x^2 + 3x$$
$$0 = x^2 + 3x - 40$$
$$0 = (x + 8)(x - 5)$$

$$x + 8 = 0 \quad \text{or} \quad x - 5 = 0$$
$$x = -8 \qquad\qquad x = 5$$

Since dimensions cannot be negative, eliminate -8.
Therefore, height is 5 feet and base is 5 + 3 = 8 feet.

Example 3.   A boat travels downstream a distance of 20 miles in the same time it takes to travel upstream a distance of 16 miles. If the current has a speed of 4 mph, find the speed of the boat in still water.

Solution: Let x = rate in still water
So x - 4 = rate upstream
    x + 4 = rate downstream

	Distance	Rate	Time	
upstream	16	x - 4	$\frac{16}{x-4}$	Used
downstream	20	x + 4	$\frac{20}{x+4}$	Time = $\frac{distance}{rate}$

Time upstream = Time Downstream

$$\frac{16}{x - 4} = \frac{20}{x + 4}$$

$$16(x + 4) = 20(x - 4)$$
$$16x + 64 = 20x - 80$$
$$-4x + 64 = -80$$
$$-4x = -144$$
$$x = 36$$

So, the speed of the boat is 36 mph

Summary:

> Problem in which two or more machines or people work together to complete a certain task are sometimes referred to as **work problems**. We represent the total amount of work done by the number one which represents one whole job completed.
>
> Part of task done by first person or machine + part of task done by second person or machine = 1 (one whole task completed
>
> To determine the part of the task completed by each person or machine, we use the formula:
>
> **Part of task completed = rate·time**

Example 4. An older pump can empty a swimming pool in 60 hours. A newer pump require 40 hours to empty the swimming pool. If both pumps are used together, how long will it take to empty the swimming pool?

Solution. Set up a table as follows:

	Rate of work	Time worked	Part of Task
Old pump	$\frac{1}{60}$	t	$\frac{t}{60}$
New pump	$\frac{1}{40}$	t	$\frac{t}{40}$

Part of pool + part of pool emptied = 1 (entire pool emptied)
emptied by   by new pump
old pump

$$\frac{t}{60} + \frac{t}{40} = 1$$

$$120\left(\frac{t}{60} + \frac{t}{40}\right) = (120)(1)$$

$$2t + 3t = 120$$
$$5t = 120$$
$$t = 24$$

It takes 24 hours to empty the pool if both pumps work together.

Exercise Set 6.7

For each problem (a) write an equation that can be used to solve the problem (b) solve the problem.

1. The base of a triangle is 5 meters longer than its height. Find the base and height of the triangle if the area is 7 square meters.

2. The height of triangle is 3 less than twice its base. Find the base and height of the triangle if the area is 27 square units.

3. One number is twice a second number. Find the number if the sum of the reciprocals is 12.

4. The numerator of fraction $\frac{2}{3}$ is increased by an amount so that the value of the resulting number is 5. Find the amount that the numerator was increased.

5. The speed of a boat in still water is 20 mph. The boat traveled 75 miles down a river in the same amount of time as it traveled 45 miles up the river. Find the rate of the river's current.

6. An express train travels 300 miles in the same amount of time that a freight train travels 180 miles. The rate of the express train is 20 mph faster than the freight train. Find the rate of each train.

7. A car and a bus leave from the same point at the same time. The car average 50 mph on the trip and then low average 40 mph on the same trip. If the car arrive at its destination 1.5 hours before the bus, find the distance traveled.

8. If one person can mow a lawn in 20 minutes and a second person requires 30 minutes to mow it, how long would it take both people to mow the lawn when they work together?

9. A pipe can fill a swimming pool three times faster than a second pipe. With both pipes working together, they can fill the pool in 9 hrs. How long does it takes for each of the pipes working separately to fill the pool?

10. A mason can construct a wall in 10 hrs. With a helper, the task would takes 6 hrs. How long would it take the helper working alone to construct the wall?

Answers to Exercise Set 6.7

1. height = 2, base = 7
2. base = 6, height = 9
3. one number = $\frac{1}{4}$

    other number = $\frac{1}{8}$
4. the amount = 13
5. rate of current = 5 mph
6. freight train = 30 mph
   express train = 50 mph
7. distance = 300 miles
8. 12 minutes
9. 12 hrs, 36 hrs.
10. 15 hours

Chapter 6 Practice Test

Perform the operations indicated:

1. $\dfrac{2x^2y}{7z^3} \cdot \dfrac{21z^4}{8x^3y^5}$

2. $\dfrac{x^5y^3}{x^2 - x - 6} \cdot \dfrac{x^2 - 9}{x^2y^4}$

3. $\dfrac{x^2 + 3x + 2}{x^2 + 5x + 4} \div \dfrac{x^2 - x - 6}{x^2 + 2x - 15}$

4. $\dfrac{x^2 - x - 56}{x^2 + 8x + 7} \div \dfrac{x^2 - 13x + 40}{x^2 - 4x - 5}$

5. $\dfrac{2x - 5}{3x} + \dfrac{x + 5}{3x}$

6. $\dfrac{2x}{x^2 + 3x - 10} - \dfrac{4}{x^2 + 3x - 10}$

7. $\dfrac{3}{a^2} - \dfrac{5}{a^3}$

8. $5 + \dfrac{2a}{a - 5}$

9. $\dfrac{3}{x^2 - 4} + \dfrac{x}{(x - 2)^2}$

Simplify each expression.

10. $\dfrac{5 - \dfrac{3}{4}}{6 + \dfrac{1}{8}}$

11. $\dfrac{a + \dfrac{1}{a^2}}{\dfrac{1}{2a}}$

12. $\dfrac{2x}{3} + \dfrac{4x}{5} = 4$

13. $\dfrac{x}{x + 4} = \dfrac{11}{x^2 - 16} + 2$

Solve the problems.

14. A small plane can fly at 110 mph in calm air. Flying with the wind, the plane can fly 260 miles in the same amount of time it can fly 180 miles against the wind. Find the rate of the wind.

15. One pipe can fill a tank in 9 hours, while a second pipe requires 18 hours to fill the tank. How long would it take both pipes working together to fill the tank?

Answers to Chapter 6 Practice Test

1. $\dfrac{3z}{4xy^4}$

2. $\dfrac{x^3(x + 3)}{y(x + 2)}$

3. $\dfrac{x + 5}{x + 4}$

4. 1

5. 1

6. $\dfrac{2}{x + 5}$

7. $\dfrac{3a - 5}{a^3}$

8. $\dfrac{7a - 25}{a - 5}$

9. $\dfrac{x^2 + 5x - 6}{(x - 2)^2(x + 2)}$

10. $\dfrac{37}{49}$ 11. $\dfrac{2a^3 + 2}{a}$ 12. $\dfrac{30}{11}$

13. −7, 3   14. 20 mph   15. 6 hours

7.1 Pie, Bar and Line Graphs

Summary:

> **Pie Graphs**
>
> 1. A **pie graph**, also called a **circle graph**, displays information using a circle.
>
> 2. The circle is divided into pieces called **sectors**.

Example 1. Study the pie graph below. Assume that there is a total of 100,000 deaths in the 25-44 male age group.

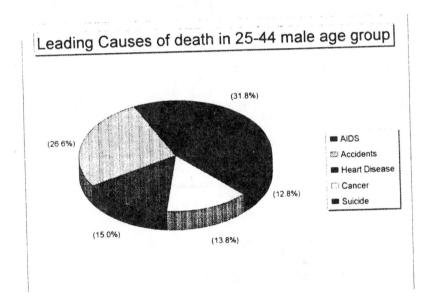

a) What should the sum of the amounts in the five sectors equal?

b). Approximately how many deaths in this age group are caused by accidents ?

c) Approximately how many deaths in this age group are caused by cancer?

Solution. a) Since the circle represents one whole or 100%, the sum of the five sectors should equal 1.00 (100%).

Summary:

> **Bar Graphs**
>
> 1. A typical bar graph indicates amounts on the vertical axis.
>
> 2. The horizontal axis may indicate additional information such as time (in years or months, usually).
>
> 3. Bar graphs can be used to indicate trends.
>
> 4. The vertical axis of a bar graph should start at zero to correctly reflect the information.

Example 2. Study the bar graph which indicates the estimated construction costs of Denver's new airport.

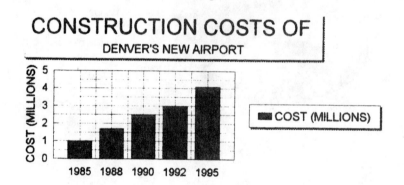

a) What were the construction costs of the airport in 1985?

b) How much more were the construction costs in 1995 than in 1985?

c) What is the percentage change in construction costs from 1985 to 1995 ?

Solution. a) Find the bar for the year 1985. This is the first bar

b) First, find the estimated construction costs in 1995. Locate the year 1995 on the horizontal axis. Extend the top of the bar until it intersects the vertical axis. The point is approximately midway between 4.0 and 4.5. So, we shall call the value 4.25. Find the difference between 4.25 and 1.0, which is 3.25. Thus, estimated construction costs were 3.25 billion dollars higher in 1995 than in 1985.

c) The percentage change can be calculated as the

$$\frac{\text{amount of change}}{\text{original amount}} \times 100\%$$

The amount of change is 3.25. The original amount is 1.0. Thus, the percentage change is $\frac{3.25}{1.00} \times 100\%$, or 325 %.

This is equivalent to saying that estimated construction costs of Denver's airport was 3.25 times higher in 1995 than in 1985.

Summary:

> **Line Graphs**
>
> 1. A **line graph** generally has amounts indicated on the vertical axis, and some measure of time, such as years or months, listed on the horizontal axis.
>
> 2. **Line graphs** can be used to predict trends in data.

Example 3. Study the line graph which indicates the annual costs of regulation of American companies (in 1991 dollars).

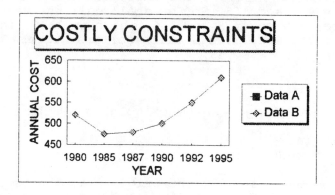

a) What type of trend does the line graph show from the years 1980 to 1985 ?

b) What type of trend does the line graph show from the years 1988 to 1995 ?

Solution. a) In 1980, the annual costs of governmental regulation was slightly more than $500 billion. In 1985, the annual costs of regulations dropped to slightly more than $450 billion. There was a decreasing trend in the costs of governmental regulation from the years 1985 to 1995.

b) From the years of 1988 to 1995, there was a continual increasing trend in the costs of regulations.

Exercise set 7.1

1. Study the bar graph below and answer the following questions.
a) What is the overall trend in average number of drinks per week and grade average ?

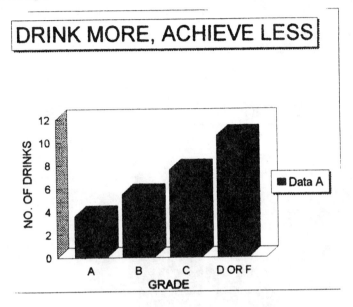

b) What is the percentage increase in the average number of drinks per week for those students whose grade average is D or F versus those students whose grade average is A ?

c) By how much does the average of drinks per week for C students exceed the average number of drinks per week for B students?

2. Study the line graph below and answer the following questions.

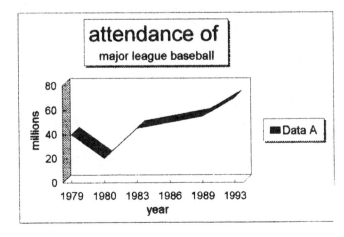

a) What is the overall trend in attendance of major league baseball games since 1980 ?

b) During what two years was the rate of growth in attendance the fastest ?

3. Consider the bar graph depicting the average major league baseball team operating costs.

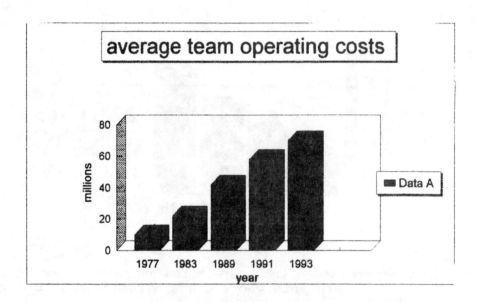

a) Approximately what was the change in operating costs from 1977 to 1993 ?

b) What is the overall trend in operating costs?

c) What is the percentage increase in operating costs from 1983 to 1989 ?

4. Study the bar graph depicting Professor X's grade distribution for Calculus I.

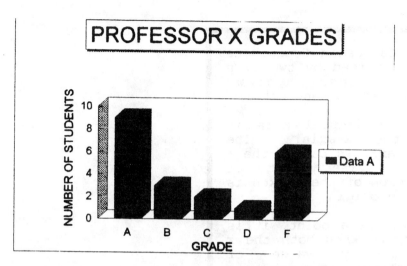

a) If all students in Professor X's class are depicted in the bar graph, how many students are enrolled in this class ?

b) What percentage of the students earned an A grade ?

c) What percentage of the students passed the class (pass means a grade of C or higher)

d) How many more A's than F's were earned ?

Answers to Exercise set 7.1

1 a). As the average number of drinks per week increases, the grade point average decreases.

b) 194.4 % increase   c). 2.1 drinks

2. a). overall increasing trend. As the year goes up, so does the attendance.

b) from 1980 - 1981 and from 1992 to 1993

3. a) approximately $60 million.   b) operating costs have risen steadily from 1977 to 1993.   c) approximately 91%

4. a) 21 students   b). 42.8%   c) 66.66%   d) 3

Section 7.2 **Graphing Linear Equations**

Summary:

> 1. The Cartesian Coordinate System is formed by two axes (number lines) drawn perpendicular to each other.
>
> 2. The horizontal axis is called the **x-axis**. The vertical axis is called the **y-axis**. The point of intersection of the 2 axes is called the **origin**.
>
> 3. To locate a point, it is necessary to know both the x value and y value or **coordinates** of the point.
>
> 4. A point is specified by an **ordered pair**, (a, b). In the ordered pair (a, b) the x coordinate is listed first, and the y coordinate is second.

Example 1.   Plot (-3, 5), (2, 4), (0, 3), (3, 0)

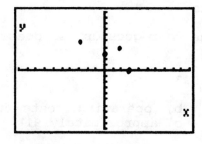

Example 2.   List ordered pairs for each point shown:

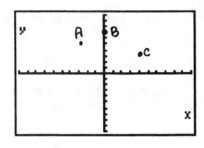

Solution. A(-3,5) , B(0,7)
            and C(4,3)

Summary:

> 1. A **linear equation in two variables** is an equation that can be put in the form
> $ax + by = c$ where a, b, c are real numbers.
>
> 2. Equations of the form **ax + by = c** will be straight lines when graphed.
>
> 3. A **graph** is an illustration of a set of points which satisfy the equation.

Example 3.   Given $y = 2x + 1$, solutions of this equation will be ordered pairs which satisfy this equation. For example, if $x = 1$, then $y = 2(1) + 1 = 3$ so (1, 3) is a solution. If $x = -2$, then $y = 2(-2) + 1$ or $y = -4 + 1 = -3$ so (-2, -3) is also a solution. If $x = 0$, then $y = 2(0) + 1 = 1$ so (0, 1) is the third solution. If we plot the 3 ordered pairs (1, 3), (-2,-3), and (0, 1), we obtain the graph shown:

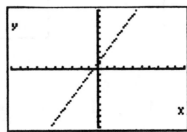

The 3 points lie on the same line. Countless other points also lie on this line.

Example 4. Determine which of the following ordered pairs satisfy $x - y = 1$.

    a) (0, 1)   b) (2, 3)   c) (-3, -4)

a) We substitute the ordered pair (0, 1) into $x - y = 1$ to obtain $0 - 1 = 1$ or $-1 = 1$ which is false. Thus (0, 1) does not satisfy the equation.

b) Substitute (2, 3) into $x - y = 1$ to obtain $2 - 3 = 1$. $-1 = 1$ which is false. Thus, (2, 3) is not a solution.

c) Substitute (4, 3) into $x - y = 1$. Thus $4 - 3 = 1$ is true so (4, 3) satisfy the equation.

Example 5. Determine if the three points are collinear.

    a) (0, 1), (2, 5), (-2, -3)
    b) (0, 3), (1, 2), (3, 0)
    c) (0, 0), (2, 8), (3, 9)

If 2 points are collinear, then that means they lie on the same line. We will plot each set of points on the grid to determine if they lie on the same line.

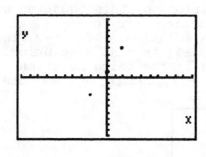
a)

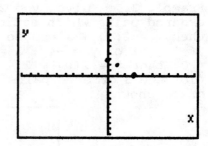
b)

c) 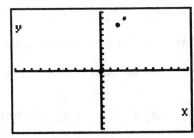 Solution. The points in a) and b) are collinear. The points in c) are **not** collinear

Exercise Set 7.2
Indicate the quadrant in which each of the points belong:

1. (3, 6)   2. (-3, 5)   3. (-3, - 5)

4. (100, 93)   5. (5, -100)

6. List the ordered pairs corresponding to the points graphed.

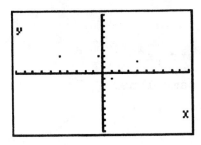

7. Plot the points. Then determine if they are collinear.

   a) (0, 3), (1, 2), (3, 0)
   b) (1, 3), (-2, 1), (0, 5)
   c) (-2, -8), (2, 4), (-3, -10)

8. Determine if the ordered pairs satisfy the equation

   $y = \frac{2}{3}x - 2$

   a) (-3, -4)   b) (0, -2)   c) (6, 0)

   d) (3, 0)   e) (9, 3)

9. Given $y = 2x + 1$, complete the ordered pairs so that each ordered pair solves the equation.
   (-3, ___)   (-1, ___)   (0, ___)   (2, ___)   (3, ___)

(-3, ___) (-1, ___) (0, ___) (2, ___) (3, ___)

10. How many points are needed to graph a linear equation?

    b) Why is it preferable to use three or more points when graphing a linear equation?

11. What does it mean when one says "three points are collinear"?

Answers to Exercise 7.2

1. I 2. II 3. III 4. I

5. IV 6. (5, 0), (-5, 3), (4, 2), (1, -1), $\left(-\frac{1}{2}, 3\right)$

7. a) yes 8. a) yes 9. -5, -1, 1, 5, 7
   b) yes   b) yes
   c) no   c) no
              d) yes
              e) no

10. a) two b) to catch errors

11. The 3 points all lie on the same line.

Section 7.3   **Graphing Linear Equations by plotting points**

Summary:

> **Graphing Linear Equations by Plotting Points**
>
> 1. Solve the linear equation for the variable y. That is, get the variable y by itself on the left side of the equal sign.
>
> 2. Select a value for the varialbe x. Substitute this value in the equation for x and find the corresponding value of y. Record the ordered pair (x, y).
>
> 3. Repeat step 2 with two different values of x. This will give you two additional ordered pairs.
>
> 4. Plot the three ordered pairs. The three points should be collinear. If they are not collinear, recheck your work for mistakes.
>
> 5. With a straight-edge, draw a straight line through the three points. Draw arrow heads on each end of the line segment to show that the line continues indefinitely in both directions.

Example 1.   Graph $y = -2x + 5$

    1. Equation already solved for y

x	$y = -2x + 5$	ordered pair
-1	$y = -2(-1) + 5 = 7$	(-1, 7)
0	$y = -2(0) + 5$	(0, 5)
3	$y = -2(3) + 5$	(3, -1)

    2. Plot (-1, 7), (0, 5), (3, -1)

Example 2. $3x - 4y = 12$

1. $-4y = -3x + 12$

   $y = \dfrac{-3x + 12}{-4} = \dfrac{-3}{-4}x + \dfrac{12}{-4} = \dfrac{3}{4}x - 3$

2. Select values of x divisible by 4 since coefficient of x is $\dfrac{3}{4}$

x	$y = \dfrac{3}{4}x - 3$	ordered pair
-4	$y = \dfrac{3}{4} \cdot \dfrac{(-4)}{1} - 3 = -3 - 3 = 0$ 	(-6, 0)
0	$y = \dfrac{3}{4}(0) - 3 = 0 - 3 = -3$	(0, -3)
4	$y = \dfrac{3}{4}(4) - 3 = 3 - 3 = 0$	(4, 0)

3. Connect points with a straight line using a straight edge.

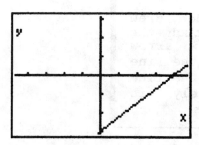

Summary:

> **Graphing Linear Equation Using the x and y Intercepts**
>
> 1. Find the y intercept by setting x in the given equation equal to 0 and finding the corresponding value of y.
>
> 2. Find the x intercept by setting y in the given equation equal to 0 and finding the corresponding value of x.
>
> 3. Determine a check point by selecting a nonzero value for x and finding the corresponding value of y.
>
> 4. Plot the y intercept (where the graph crosses the y axis), the x intercept (where the graph crosses the x axis), and the check point. The three points should be linear. If not, recheck your work.
>
> 5. Using a straight-edge, draw a straight line through the three points. Draw arrow head at both ends of the line to show that the line continues indefinitely in both directions.

Example 3.   Graph $2y = 5x + 10$ using x and y intercepts

1. Find y intercept by setting $x = 0$:
$2y = 5(0) + 10$
$2y = 0 + 10$
$2y = 10$
$y = 5$

Thus, the y intercept is $(0, 5)$

2. Find the x intercept by setting y = 0
   $$2(0) = 5x + 10$$
   $$0 = 5x + 10$$
   $$-10 = 5x$$
   $$\frac{-10}{5} = \frac{5}{5}x$$
   $$-2 = x$$

   So x intercept is (-2, 0)

3. Select a non-zero value of x and find the corresponding value of y and make sure it is collinear with the x and y intercepts.
   Let x = 2   Then 2y = 5(2) + 10
   $$2y = 10 + 10$$
   $$2y = 20$$
   $$y = 10$$

   So (2, 10) is a third point.

   Now, plot all 3 points and make sure they are collinear.

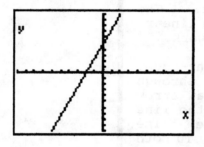

Example 4.   Graph y = -4

1. Write the equation as y = 0x - 4

2. Make a table of values:

x	y = 0x - 4	ordered pair
-2	y = 0(-2) - 4 = -4	(-2, -4)
0	y = 0(0) - 4 = -4	(0, -4)
2	y = 0(2) - 4 = -4	(2, -4)

As you can see, no matter what x is, y is always equal to -4.

3. Plot the ordered pairs.

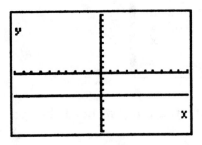

Example 5. Graph the equation x = 5

1. x = 5 + 0y

2. Instead of choosing x values first, we will choose arbitrary values of y, first.
If y = -3 then x = 5 + 0(-3) = 5
So (5, -3) is a point.

If y = 0 then x = 5 + 0(0) = 5
So (5, 0) is a second point.

If y = 3 then x = 5 + 0(3) = 5
So (5, 3) is a third point.

As you can see, no matter what value is chosen for y, x is always 5.

3. Graph

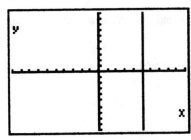

Summary:

> 1. The graph of any equation of the form $y = b$ is a horizontal line whose y intercept is b.
>
> 2. The graph of any equation of the form $x = a$ is a vertical line whose x intercept is a.

Example 6.   The cost of joining a health spa is $150 plus $24 per month.

a) Write an equation for the cost, C, in terms of number of months, n. Make a table of values:

n	C
0	150
1	150 + 1(24)
2	150 + 2(24)
3	150 + 10(24)

From the pattern of the table, we have $C = 150 + 24n$.

b) Graph the cost for number of months ranging from 0 to 12. The equation is linear. We will find 3 ordered pairs which solve the equation.

n	C = 150 + 24n	ordered pair
0	C = 150 + 24(0)	(0, 150)
5	C = 150 + 24(5)	(5, 270)
10	C = 150 + 24(10)	(10, 390)

To graph this equation, we must choose a suitable vertical scale.

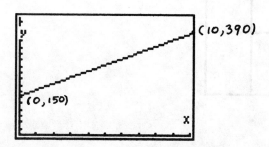

c) Use the graph to estimate the cost for using the spa for 7 months.

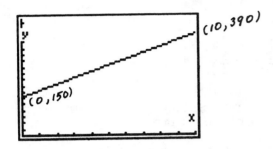

Find the number 7 on the month's axis. Draw a vertical line up to where the graph is intersected. Read the y value from the graph. It is approximately $320.00. You can obtain the exact value by substituting n = 7 into the equation and finding C.

C = 150 + 24(7) = 150 + 168 = 318.

Exercise set 7.3

Find the missing coordinate in the solutions for 3x + y = 9

1. (2,)   2. (, 3)   3. (-4,)

4. (0,)   5 (, 0)

Find the missing coordinate in the solutions for 2x - 7y = 14.

6. (0,)   7. (, 0)   8. (-3,)

9. (, 5)

Graph each equation

10. x = -6   11. y = 5   12. y = -3

13. x = 2

Graph each equation by plotting points. Plot at least 3 points for each graph.

14. y = -3x + 2   15. y = -x + 4

16. y = 5x + 3   17. y = $\frac{2}{3}$x + 1

18. $-2x + 3y = 6$ 19. $-3x - y = 2$

Graph using the x and y intercepts.

20. $4x - y = 16$ 21. $2x - 3y = 12$

22. $\frac{1}{3}x - \frac{3}{4}y = 60$ 23. $\frac{2}{3}y = \frac{5}{4}x - 120$

24. Using the formula I = prt, if p = $5000.00, r = 4%, write an equation for simple interest in terms of time, t.

b) Graph the equations for times of 0 to 20 years inclusive.

c) What is the simple interest for a time of 8 years?

Answers to exercise set 7.3

1. 3   2. 2   3. 21   4. 9   5. 3

6. -2   7. 7   8. $\frac{-20}{7}$ 9. $\frac{49}{2}$

10.

11.

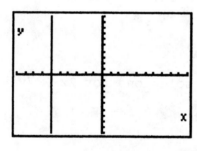

12.

13.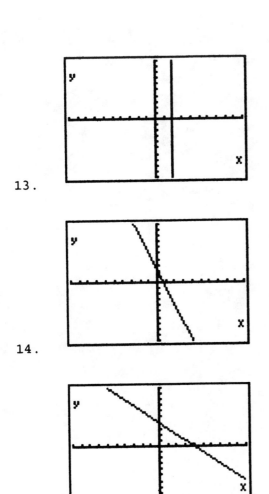

14.

15.

16.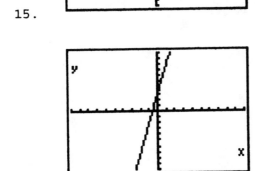

17.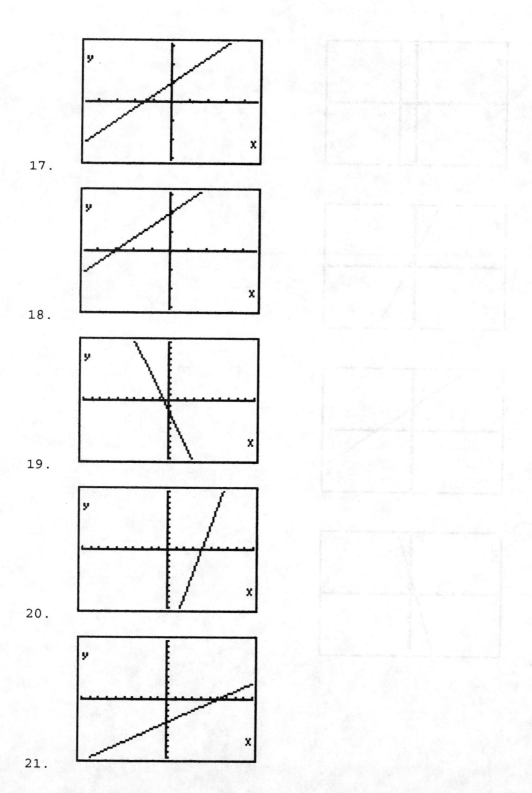

18.

19.

20.

21.

22.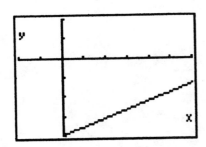

23.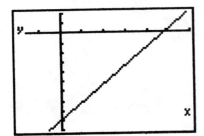

24. I = 5000(.04)t   b)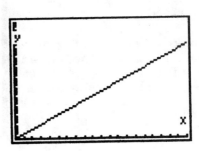

c) $1600

7.4 Slope of a Line

Summary:

> 1. The slope of a line measures the **steepness** of the line.
>
> 2. The **slope of a line** is a ratio of the vertical change to the horizontal change between **any two** arbitrary points on the line.
>
> 3. The definition of the slope of a line through (x_1, y_1) and (x_2, y_2) is as follows:
>
> $$\text{Slope} = \frac{y_2 - y_1}{x_2 - x_1}$$
>
> 4. It makes no difference which two points are selected when you find the slope. It also makes no difference which point you label as (x_1, y_1) or (x_2, y_2).

Example 1. Find the slope of the line containing the points (-6, 5) and (3, -4).

Solution. Slope $= \dfrac{y_2 - y_1}{x_2 - x_1} = \dfrac{5 - (-4)}{-6 - 3} = \dfrac{9}{-9} = -1$

or slope $= \dfrac{y_1 - y_2}{x_1 - x_2} = \dfrac{5 - (-4)}{-6 - 3} = \dfrac{9}{-9} = -1$.

Notice that you can reverse the order of subtraction **provided** you do it for **both the change in y as well as the change in x.**

Summary:

1. A straight line has **positive** slope if the value of y increases as the value of x increases.

2. A straight line has **negative** slope if the value of y decreases as x increases.

Example 2. Consider the line graphed below:

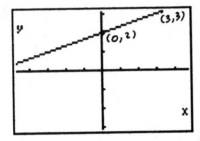

a) using the 2 given points (0,2) and (3,3) , determine the slope of the line by observing the vertical change and the horizontal change.

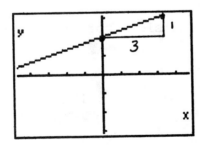

You can see that the slope must be positive since y increases as x increases. The vertical change between the two points is one unit. The horizontal change between the two points is 3 units. Thus, the slope is $\frac{1}{3}$.

b) Find the slope of the line using the two points.

Using (0,2) and (3,3) and the slope formula: $m = \frac{y_2 - y_1}{x_2 - x_1}$

Example 3. Find the slope of the line below by observing the vertical change and horizontal change between the two points shown.

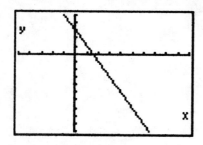

First, the slope is negative since y decreases as x increases. Next, observe that the vertical change between the two points is -8 and the horizontal change is 4 . Thus, the slope, m , is $\frac{-8}{4}$ = -2 .

You can verify this by using the slope formula $m = \frac{y_2 - y_1}{x_2 - x_1}$.

$m = \frac{-7 - 1}{5 - 1} = \frac{-8}{4} = -2$

Summary:

> 1. Every **horizontal** line has a slope of zero.
>
> 2. The slope of any **vertical** line is undefined.

Example 4. Graph $y = -2$. What is the slope of this line?

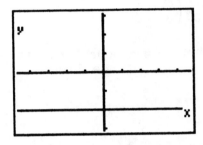

Since this line is horizontal, its slope is zero. Notice that there is **no** change in y between any two points on the line.

Example 5. Graph $x = 5$

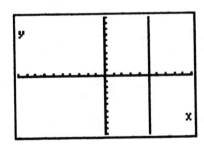

Since this line is vertical, its slope is **undefined.** This line passes through the points $(5,-1)$ and $(5,7)$. If you calculate the slope using these points, the result is $m = \frac{7 - (-1)}{5 - 5} = \frac{8}{0}$ (undefined).

Exercise set 7.4

Find the slope of the line through the given points.

1. (3,7) and (6, 16)
2. (1,4) and (3,8)
3. (3,-4) and (4,-3)
4. (0,5) and (8,-3)
5. (-7, -5) and (3,-4)
6. (4,-2) and (-6,5)
7. (5,7) and (5, -10)
8. (3, 12) and (-3, 12)
9. (5, -9) and (-3, -7)
10. $\left(\frac{1}{2}, \frac{3}{4}\right), \left(\frac{5}{8}, \frac{7}{8}\right)$

By observing the vertical and horizontal change of the line between the two points indicated, determine the slope of the line.

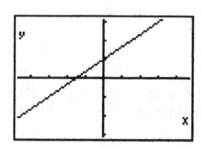
11.

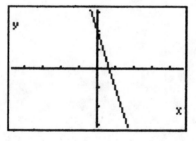
12.

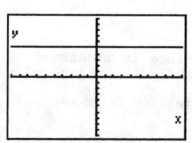
13.

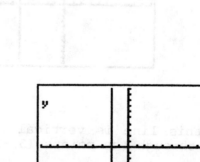
14.

Answers to exercise set 7.4

1. 3   2. 2   3. 1   4. -1   5. $\frac{1}{10}$

6. $\frac{-7}{10}$ 7. undefined   8. 0   9. $\frac{-1}{4}$

10. 1   11. $m = \frac{2}{3}$ 12. $m = -3$

13. $m = 0$ 14. undefined slope

7.5 Slope-Intercept and Point-Slope Forms of a Linear Equation

Summary:

> 1. The **slope-intercept form** of a linear equation is
> $y = mx + b$ where $m = $ **slope** and $b = $ **y-intercept**.
>
> 2. To write a linear equation in **slope-intercept form**, solve the equation for y

Example 1. Given the following equations of lines, determine the slope and y-intercept.

a) $y = -2x + 5$

Solution. Since the coefficient of x is the slope, the slope is -2 and the y-intercept is 5.

b) $y = \frac{-3}{4} x - 3.$

Solution. Here, the coefficient of x is $-\frac{3}{4}$ and the y-intercept is -3.

Example 2. Write the equation $2x - 5y = -10$ in slope-intercept form. State the slope and the y-intercept.

Solution. The equation should be solved for y.

$$2x - 5y = -10$$
$$-5y = -2x - 10$$
$$\frac{-5y}{-5} = \frac{-2x}{-5} + \frac{-10}{-5}$$
$$y = \frac{2}{5}x + 2$$

The coefficient of x is the slope, which is $\frac{2}{5}$ and the y-intercept is 2.

Summary:

> **To Graph Linear Equations Using the Slope and y-intercept**
>
> 1. Solve the equation for y to obtain $y = mx + b$.
>
> 2. Note the slope, m, and the y-intercept.
>
> 3. Plot the y-intercept on the y-axis.
>
> 4. Use the slope to find a second point on the graph.
>
> 5. Use a straight edge to draw a straight line through the two points.

Example 3. Write the equation $-2x + 5y = -15$ in slope-intercept form. Then, use the slope and y-intercept to graph this line.

Solution.

Step 1. Solve $-2x + 5y = -15$ for y

$$5y = 2x - 15$$

$$\frac{5y}{5} = \frac{2x}{5} - \frac{15}{5}$$

$$y = \frac{2}{5}x - 3$$

Step 2. Note the slope and y-intercept. The slope is $\frac{2}{5}$ and y-intercept is -3.

Step 3. Plot the y-intercept $(0, -3)$

Step 4. Use the slope to find a second point on the graph.

From the point (0, -3), move up 2 units and to the right 5 units to obtain a second point (5, -1) .

Step 5. Use a straight-edge to draw a straight line through the two points. A third point can be found by moving up 2 units and to the right 5 units from the point (5, -1). This third point (10, 1) should lie on the straight line.

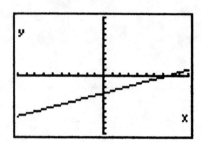

Example 4. Graph $-2x + 3y = 18$ using the slope and y-intercept.

Solution.

Step 1. Solve for y

$$3y = 2x + 18$$

$$\frac{3y}{3} = \frac{2x}{3} + \frac{18}{3}$$
$$y = \frac{2x}{3} + 6$$

Step 2. Note slope and y-intercept. Slope is $\frac{2}{3}$ and y-intercept is 6.

Step 3. Plot the y-intercept (0,6)

Step 4. Use the slope to find a second point on the graph.

From the point (0,6) , go up two units and right 3 units to obtain the point (3,8). Plot this point.

Step 5. Use a straight edge to connect the two points. Plot a third point by going up two units and right 3 units from the point (3,8) to obtain (6,10). Make sure this point lies on the straight line drawn.

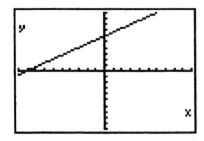

Example 5. Determine the equation of the line using the graph shown.

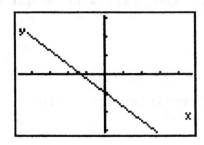

Solution. Since the y-intercept is (0, -1) and the slope is $\frac{-3}{4}$, the equation is $y = \frac{-3}{4}x - 1$

Summary:

> 1. Lines with the **same slope** but **different** y-intercepts are parallel.
>
> 2. If the slopes of the two lines are **not the same**, the lines are **not parallel**.
>
> 3. If two lines have the **same** slope and the **same** y-intercept then both equations represent the **same line**.

Example 6. Determine whether or not the following equations represent parallel lines.

$$y = -3x + 1$$
$$6x - 2y = 7$$

Solution. Write each equation in slope-intercept form. The equation $y = -3x + 1$ is already in that form.

Solve $6x - 2y = 7$ for y in terms of x

$$-2y = -6x + 7$$

$$\frac{-2y}{-2} = \frac{-6x}{-2} + \frac{7}{-2}$$

$$y = 3x - \frac{7}{2}$$

The slope of the first line is -3 but the slope of the second line is 3. So, the lines are **not parallel**.

Summary:

> **Point-slope form of a linear equation**
>
> $$y - y_1 = m(x - x_1)$$
>
> m is the slope of the line and (x_1, y_1) is a point on the line.

Example 7. Write an equation of a line which passes through (-2, 4) and has a slope of 5.

Solution.

Use the point slope form $y - y_1 = m(x - x_1)$ where $(x_1, y_1) = (-2, 4)$ and m = 5.

$$\begin{aligned} y - 4 &= 5(x - (-2)) \\ y - 4 &= 5(x + 2) \\ y - 4 &= 5x + 10 \\ y - 4 + 4 &= 5x + 10 + 4 \\ y &= 5x + 14 \end{aligned}$$

Example 8. Write an equation of the line through the points (-2,-5) and (3,10).

Solution.

Step 1. Find the slope of the line.

$$\text{Use } m = \frac{y_2 - y_1}{x_2 - x_1} = \frac{10 - (-5)}{3 - (-2)} = \frac{15}{5} = 3$$

Step 2. Use the point-slope form of the equation of a line. Use the slope of 3 found in step 1 and either point as (x_1, y_1)

$$y - y_1 = m(x - x_1)$$

$$\begin{aligned} y - (-5) &= 3(x - (-2)) \\ y + 5 &= 3(x + 2) \\ y + 5 &= 3x + 6 \\ y + 5 - 5 &= 3x + 6 - 5 \\ y &= 3x + 1 \end{aligned}$$

Summary:

> **Three methods to graph linear equations**
>
> 1. Plotting points
>
> 2. using x and y-intercepts
>
> 3. Using the slope and y-intercept.

Example 9. Graph $4x - 3y = 9$ by a) plotting points, b) using intercepts and c) using the slope and y-intercept.

Solution.

a) Get equation in slope-intercept form by solving it for y

$$4x - 3y = 9$$
$$-3y = -4x + 9$$
$$y = \frac{-4x}{-3} + \frac{9}{-3}$$
$$y = \frac{4x}{3} - 3$$

Now, make a table of ordered pairs.

                                                          Ordered pair

If $x = -3$, $y = \frac{4}{3} \cdot -3 - 3 = -7$ $(-3, -7)$

If $x = 0$, $y = \frac{4}{3} \cdot 0 - 3 = -3$ $(0, -3)$

If $x = 3$, $y = \frac{4}{3} \cdot 3 - 3 = 1$ $(3, 1)$

Notice that values of x were chosen that were divisible by 3 because of the fraction $\frac{4}{3}$ in front of the x.

Plot all three points and connect them with a straight-edge.

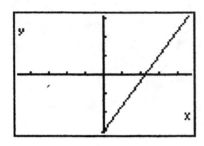

Exercise set 7.5

Determine the slope and y-intercept of the line represented by each equation. Graph the line using the slope and y-intercept.

1. $y = -2x + 3$
2. $3x - y = 5$
3. $4x - 4y = 8$
4. $4x - 5y = 20$
5. $5x - 4y = 8$

Determine the equation of each line graphed below.

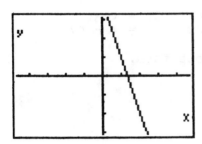

6.

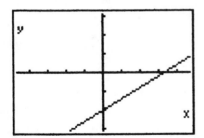

7.

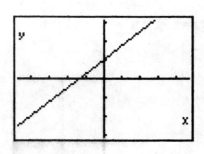
8.

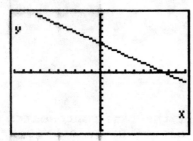
9.

Determine if the lines are parallel.

10.   y = -2x + 5   11. 3x - 8y = 24

       2y + 4x = 0   y = $\frac{-3}{8}$x+5

12.   4x - 7y = 10   13. 4x + 2y = 10
    - 8x + 14y = 12   8x   = 4 - 4y

Write the equation in slope-intercept form.

14. slope = 5 through (-2,3)

15. slope = $\frac{3}{2}$ through (-1,3)

16. slope = $-\frac{1}{3}$ through (3, -4)

17. through (3,3) and (7,7)

18. through (3,1) and (6,-1)

19. through (1,1) and (5, -15)

20. Graph the equation -4x + 2y = 6 by
    a) plotting points, b) using x and y-intercepts, c) using slope and y intercept.

Answers to exercise set 7.5

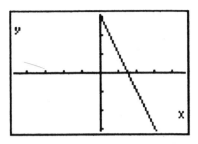

1. slope is -3, y intercept 3

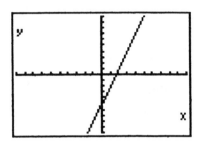

2. slope is 3, y-intercept (0,-5)

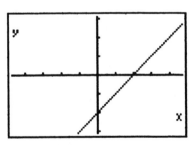

3. slope is 1, y intercept (0,-2)

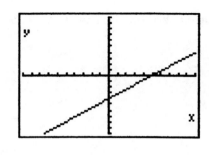

4. slope is $\frac{4}{5}$, y-intercept (0,-4)

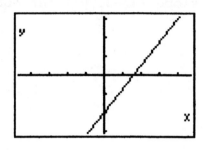

5. slope is $\frac{5}{4}$, y-intercept (0,-2)

6. $y = -3x + 4$ 7. $y = \frac{3}{5}x - 2$ 8. $y = \frac{3}{4}x + 1$

9. $y = \frac{-2}{3}x + 5$ 10. Yes   11. No   12. Yes

13. Yes   14. $y = 5x + 13$ 15. $y = \frac{3}{2}x + \frac{9}{2}$

16. $y = \frac{-1}{3}x - 3$ 17. $y = x$ 18. $y = \frac{-2}{3}x + 3$

19. $y = -4x + 5$

20. a,b,c have same graph

20.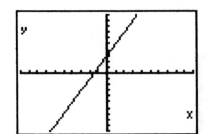

7.6 Functions

Summary:

> **Relations**
>
> 1. A **relation** is any set of ordered pairs.
>
> 2. A relation may be indicated by
>   a) any **equation** in two variables.
>   b) a **set** of ordered pairs
>   c) a **graph**

Example 1. Given $y = 2x + 3$ where $x = 1,2,3,4,5$.
This equation is a relation. We can obtain ordered pairs that satisfy this relation as follows:

	$y = 2x + 3$	ordered pair
Let $x = 1$,	$y = 2(1) + 3 = 5$	(1,5)
Let $x = 2$	$y = 2(2) + 3 = 7$	(2, 7)
Let $x = 3$	$y = 2(3) + 3 = 9$	(3, 9)
Let $x = 4$	$y = 2(4) + 3 = 11$	(4, 11)
Let $x = 5$	$y = 2(5) + 3 = 13$	(5, 13)

The set of ordered pairs { (1,5), (2,7), (3,9), (4,11), (5,13)} is a relation.

If the above five ordered pairs are plotted, the graph is a relation.

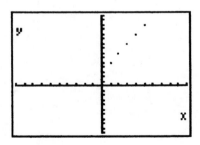

Summary:

Domain and Range

1. The **domain** of a relation is the set of **first** coordinates in the set of ordered pairs.

2. The **range** of a relation is the set of **second** coordinates in the set of ordered pairs.

Example 2. Given the set of ordered pairs from our first example, {(1,5), (2,7), (3,9), (4,11), (5,13)}, the domain is {1,2,3,4,5} and the range is {5,7,9,11,13}.

Example 3. Determine the domain and range of the relation graphed below.

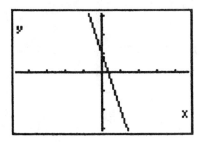

The arrowheads on the line indicate that the graph continues indefinitely. Any real number for x can be used in the domain so the domain is **R** (real numbers). Similarly, the range is **R**.

Summary:

> **Functions**
>
> 1. A **function** is a special type of relation in which no **two** ordered pairs have the **first first coordinate and a different second coordinate**.
>
> 2. A **function** may also be defined as a relation in which each element of the domain corresponds to **exactly one** element of the range.
>
> 3. The **vertical line test** is used to determine if a graph is a function.
>
> a) If a vertical line can be drawn through any part of a graph and intersects that graph **more than once**, then the graph is **not a function.**
>
> b) If a vertical line drawn through any part of a graph intersects that graph **exactly once**, then the graph **is a function.**
>
> 4. **Warning!** All functions are relations but **not all** relations are functions.

Example 4. Given the set of ordered pairs {(1,5), (2,7), (3,9), (4,11), (5,13)}, this set is a function since each value of x corresponds to exactly one value of y.

If these five ordered pairs are plotted, we can see that **any** vertical line drawn through the graph will intersect that graph **at most once**. So, the graph passes the vertical line test and is a function.

Example 5. Determine, using the vertical line test, whether or not the following graphs are functions.

a)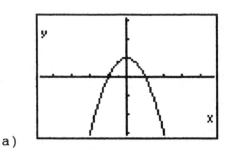
b)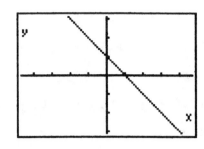

c) (graph of a circle)
d)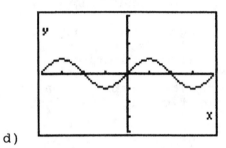

Since graphs a, b, and d all pass the vertical line test, these 3 graphs are functions while graph c is not a function.

Summary:

F(x) notation

1. The notation $y = f(x)$ is used to show that y is a function of the variable x.

2. **F(x) does not mean f times x.**

Example 6. For $f(x) = 2x^2 + 3x$, find the following:

a) $f(-3)$ b) $f(2)$ c) $f(-1)$ d) $f(t)$

Solution.

a) Since $f(x) = 2x^2 + 3x$, we can think of $f(x)$ as
$f(\ ) = 2(\ )^2 + 3(\ )$. We can substitute numbers or variables in the ( ).

So, $f(-3) = 2(-3)^2 + 3(-3) = 2(9) + -9 = 9$

b) $f(2) = 2(2)^2 + 3(2) = 2(4) + 6 = 8 + 6 = 14$

c) $f(-1) = 2(-1)^2 + 3(-1) = 2(1) - 3 = 2 - 3 = -1$

d) $f(t) = 2(t)^2 + 3(t) = 2t^2 + 3t$

Summary:

Linear Functions

1. Equations of the form $y = f(x) = ax + b$ are linear functions.

2. The graph of a linear function is a straight line.

Example 7. Graph $f(x) = -2x + 3$

Solution. Since $f(x)$ is the same as y, we have $f(x) = y = -2x + 3$.

Select arbitrary values of x and find the corresponding y values.

	$y = f(x) = -2x + 3$	Ordered pair
If $x = -3$, then	$y = f(-3) = -2(-3) + 3 = 9$	$(-3, 9)$
If $x = 0$, then	$y = f(0) = -2(0) + 3 = 3$	$(0, 3)$
If $x = 2$, then	$y = f(2) = -2(2) + 3 = -1$	$(2, -1)$

Plot the points and draw the graph, which is a straight line.

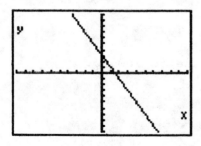

Example 8. The cost, C, of a small pizza is $5.00 plus $.75 for each additional topping (up to 5 toppings). The cost is a function of the number of toppings and is given by $C = f(n) = .75n + 5.00$, where $n = \{0, 1, 2, 3, 4, 5\}$

a) Construct a graph showing the relationship between the number of toppings and the cost.

Solution.   Select values of n; find corresponding values of C.

n	$C = f(n) = .75n + 5.00$	ordered pair
0	$C = f(0) = .75(0) + 5 = 5$	(0, 5)
1	$C = f(1) = .75(1) + 5 = 5.75$	(1, 5.75)
2	$C = f(2) = .75(2) + 5 = 6.50$	(2, 6.50)
3	$C = f(3) = .75(3) + 5 = 7.25$	(3, 7.25)
4	$C = f(4) = .75(4) + 5 = 8.00$	(4, 8.00)
5	$C = f(5) = .75(5) + 5 = 8.75$	(5, 8.75)

Plot the points and graph

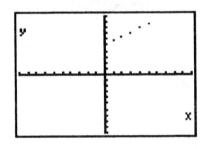

Exercise set 7.6

Determine which of the relations are also functions. Give the range and domain of each relation or function.

1. {(0,1), (2,-5), (3, -8), (4,-11)}

2. {(-2,0), (1,6), (2,8), (3,10)}

3. {(5,6), (7,2), (5,9), (6,3), (-2,3) }

The domain and range of a relation are illustrated. a) Construct a set of ordered pairs that represent the relation. b) Determine if the relation is a function.

4. Domain   Range
   -2--------> 5
   -1--------> 5
    0 --------> 5
    1 --------> 5
    2--------> 7

5. Domain   Range

6. Domain   Range

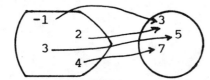

Use the vertical line test to determine if the relation is also a function.

7.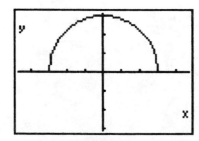

8.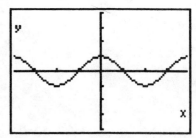

9. 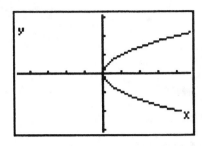 10.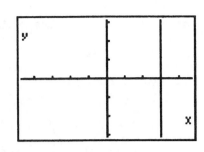

Evaluate the functions at the values indicated.

11. f(x) = 3x - 4   a) f(2)   b) f(-2)

12. f(x) = $\frac{2}{3}$x - 4   a) f(3)   b) f(-3)

13. f(x) = -2x + 7   a) f(5)   b) f(3)

Evaluate the functions at the values indicated.

14. f(x) = 2x² - 3   a) f(-2)   b) f(0)

15. f(x) = (x - 3)²   a) f(7)   b) f(-2)

16. f(x) = -2x² + 3x + 4   a) f(3)   b) f(h)

Graph each function.

17. f(x) = -2x + 3   18. f(x) = 3x - 2

19. f(x) = -4x + 2

20. A salesperson earns $200.00 per week plus 20% commission on total weekly sales. The salary, s, of the salesperson is a function of total weekly sales, x, as follows: s = 200 + .20x.

a) Draw a graph illustrating the sales for sales up to and including $5,000.

b) From the graph, estimate the salesperson's salary if total sales are $3,000.

21. What is a relation?

22. What is a function?

Answers to Exercise set 7.6

1. a function
2. a function
3. not a function
4. {(-2,5), (-1,5), (0,5), (1,5), (2,7)} a function
5. {(-2,4), (3,5), (3,7), (4,9)} not a function
6. {(-1,3), (2,3), (3,5), (4,7)} a function
7. function
8. function
9. not a function
10. not a function
11. a) 2   b) -10
12. a) -2   b) -6
13. a) -3   b) 1
14. a) 5   b) -3
15. a) 16   b) 25
16. a) -5   b) $-2h^2 + 3h + 4$

17.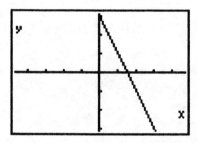

18.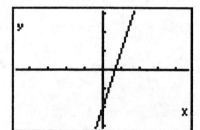

19.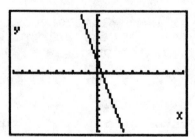

20. a) 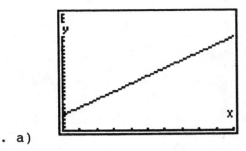 b) $800.00

21. A relation is any set of ordered pairs.

22. A relation in which no two ordered pairs have the same first coordinate and a different second coordinate.

7.7 Graphing Linear Inequalities

Summary:

To Graph a Linear Inequality

1. Replace the inequality symbol with an equal sign.

2. Draw the graph of the equation in step 1. If the original inequality contained the symbol $\leq$ or $\geq$ draw the graph using a solid line. If the original inequality contained the symbol $>$ or $<$ draw the graph using a dashed line.

3. Select **any point** not on the line and determine if this point is a solution to the original inequality. If the selected point is a solution, shade the region on the side of the line containing this point. If the selected point does not satisfy the inequality, shade the region on the side of the line not containing this point.

Example 1. Graph $y \geq -2x + 3$

Solution. Step 1. Replace the inequality symbol with an equals symbol.

$$y = -2x + 3$$

Step 2. Graph the line $y = -2x + 3$. This is a line with slope equal to -2 and y-intercept (0,3). The line is solid since the original inequality symbol is $\geq$.

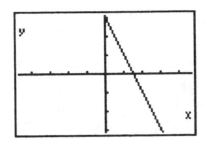

Step 3. Select **any point** not on the line. We will choose the origin since (0,0) is **not** on the line

$y \geq -2x + 3$
$0 \geq -2(0) + 3$
$0 \geq 0 + 3$
$0 \geq 3$ is **false**

We will shade the region **not** containing the origin.

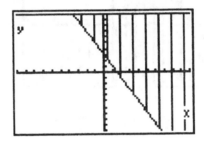

Example 2.   Graph $y < \frac{-3}{4}x$

Step 1. Replace $y < \frac{-3}{4}x$ with $y = \frac{-3}{4}x$

Step 2. Graph $y = \frac{-3}{4}x$. We will use a dashed line since the original inequality symbol is <.

Step 3. Choose a point not on the line We will choose (1,1).

$$y < \frac{-3}{4}x$$

$$1 < \frac{-3}{4}(1)$$

$$1 < \frac{-3}{4} \quad \text{is false.}$$

Shade the region **not including** the point (1,1).

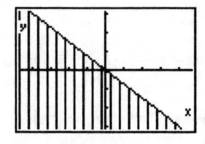

Example 3. Graph $-2x + 5y \leq 10$

Step 1. Solve for y first

$$-2x + 5y = 10$$
$$5y = 2x + 10$$
$$y = \frac{2}{5}x + 2$$

Step 2. Graph the line $y = \frac{2}{5}x+2$

The line will be solid since the original inequality symbol is ≤.

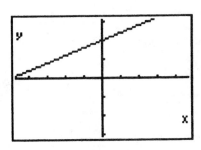

Step 3. Choose a point **not** on the line. The origin is a good choice. Determine if this point solves the inequality or not.

$$-2x + 5y \leq 10$$
$$-2(0) + 5(0) \leq 10$$
$$0 + 0 \leq 10 \text{ True}$$

Shade the region below the line containing the origin.

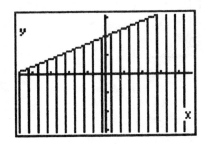

Exercise set 7.7

Graph each inequality.

1. $x > 4$

2. $y \leq -5$

3. $y \geq x$

4. $y \leq 4x$

5. $y > 2x - 4$

6. $y \leq \frac{-x}{2} + 3$

7. $2y - x < 8$ 8. $3x - 2y \leq 6$

9. $5x + 4y \geq 20$ 10. $-4x + 3y < 12$

Answers to exercise set 7.7

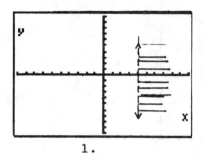
1.

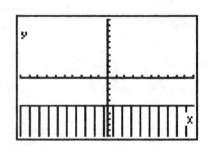
2.

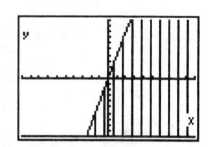
3.   4.

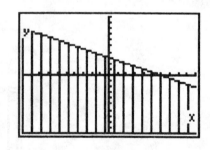
5.   6.

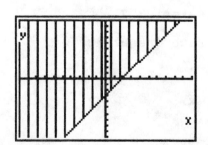
7.   8.

9. 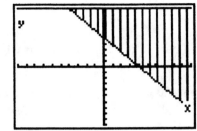   10.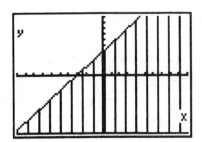

Chapter 7 Practice Test

1. Which of the following ordered pairs satisfy the equation $5x - 10y = 20$?

    a) (0,2)   b) (4,0)   c) (0,-2)   d) (-2,3)
    e) (-2,-3)

2. Find the slope of the line through the two points (-4,5) and (5,9).

3. Write an equation for the graph in the accompanying figure:

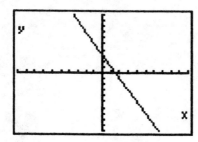

4. Write an equation in slope-intercept form of the line with a slope of -4 passing through (2,5) .

5. Write in slope-intercept form an equation of the line passing through (3,1) and (6,-1) .

6. Determine if the following equations represent parallel lines. Explain your answer.

$$y = \frac{-3}{4}x + 5$$

$$3x + 4y = 12$$

7. Find the slope and y-intercept of the line whose equation is $-5x + 3y = 15$.

Graph the following equations.

8. $x = 7$ 9. $y = -3x + 4$ 10. $-2x + 3y = 9$

11. $y = -7$

12. Determine whether the relation below is a function. Explain your answer.

{(-2,3), (-1,4), (0,3), (1,4), (2,7), (3,9) }

13. Give the domain and range of the relation
S = {(-2,3), (-1,4), (0,3), (1,4), (2,7), (3,9) }

14. Determine if the following graphs are functions. Explain your answer.

a).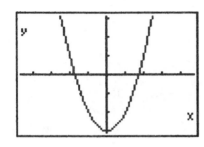

b)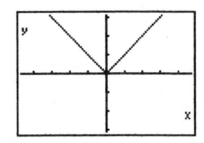

c).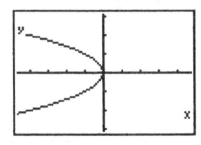

d)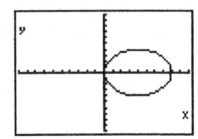

15. Given $f(x) = -2x^2 + 3x + 4$, find the following.

a) $f(-2)$ b) $f(3)$ c) $f(0)$

Graph the following equations or inequalities.

16. $f(x) = \frac{-3}{5}x + 1$

17. $y \geq -2x + 5$

18. $y < 4x + 1$

19. $x \geq 5$

Chapter 7 Practice test answers

1. b,c,e

2. $m = \frac{4}{9}$

3. $y = -2x + 3$

4. $y = -4x + 13$

5. $y = \frac{-2}{3}x + 3$

6. Yes, they are parallel because their slopes are the same. Each slope is equal to $\frac{-3}{4}$.

7. Slope is $\frac{5}{3}$; y-intercept is (0,5)

8.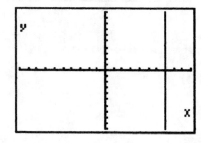

9.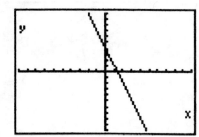

10. 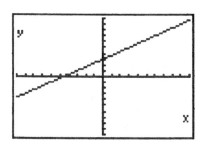   11.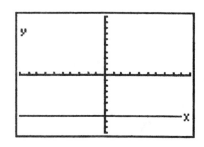

12. The relation is a function. No two ordered pairs have the same first element corresponding to different second elements.

13. Domain = { -2, -1, 0, 1, 2, 3}   Range = {3, 4, 7, 9}

14. a, b are functions since they pass the vertical line test.

15. a) -10   b) -5   c) 4   d) $-2t^2 + 3t + 4$

16. 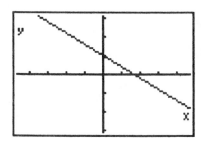   17.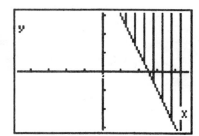

18. 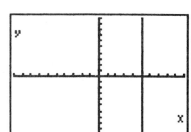   19.

8.1 Solving Systems of Equations Graphically

Summary

> A **system of linear equations** or **simultaneous linear equations** consist of two or more linear equations. The **solution to a system of equations** is the ordered pair or pairs that will satisfy all equations.

Example 1 Determine which of the following ordered pairs satisfy the system of equations.

$$3x + 4y = 18$$
$$2x - y = 1$$

a. (6,0)   b. (3,5)   c. (2,3)

a. Substitute 6 for x and 0 for y in both equations.

```
3x + 4y      = 18              2x - y    = 1
3(6) + 4(0)  = 18              2(6) - 0  = 1
18 + 0       = 18              12 - 0    = 1
18           = 18  True        12        = 1   False
```

Since (6,0) does not satisfy both equations, it is not a solution to the system of equations.

b. Substitute 3 for x and 5 for y in both equations.

```
3x + 4y      = 18              2x - y    = 1
3(3) + 4(5)  = 18              2(3) - 5  = 1
9 + 20       = 18
19           = 18  False       6 - 5     = 1
```

Since (3,5) does not satisfy both equations, it is not a solution of the system of equations.

c. Substitute 2 for x and 3 for y in both equations.

```
3x + 4y      = 18              2x - y    = 1
3(2) + 4(3)  = 18              2(2) - 3  = 1
6 + 12       = 18              4 - 3     = 1
18           = 18  True                  1 = 1
```

Since (2,3) satisfies both equations, it is a solution to the system of equations.

Summary

> The **solution to a system of linear equations** is the ordered pair (or pairs) common to all lines in the system when the lines are graphed.
>
> If one point is common to both equations, the system of linear equations is **consistent**. The linear equations have different slopes.
>
> If no points are common to both equations, the system of linear equations is **inconsistent**. The linear equations have the same slope but different y-intercept.
>
> If an infinite number of points are common to both equations, the system of linear equations is **dependent**. The linear equations have the same slope and the same y-intercept.

Example 2. Determine if the system is consistent, inconsistent, or dependent.

$$2x - 3y = 8$$
$$-4x + 6y = -16$$

Solution: Write each in slope-intercept form and then compare the slopes and y-intercepts.

$$2x - 3y = 8$$
$$-3y = -2x + 8$$
$$y = \frac{2}{3}x - \frac{8}{3}$$

$$-4x + 6y = -16$$
$$6y = 4x - 16$$
$$y = \frac{2}{3}x - \frac{8}{3}$$

Since the equations have the same slopes and he same y-intercept, the system is dependent.

Summary

> To obtain the solution to a system of equations graphically, graph each equation and determine the point or points of intersection

Example 3 Solve the following systems of equations graphically

(a) $3x - 2y = 6$
    $3x + 2y = 6$

Solution: Solve each equation for y and graph using slope and y-intercept.

$3x - 2y = 6$ $3x + 2y = 6$
$\quad -2y = -3x + 6$ $\quad 2y = -3x + 6$

$\quad y = \frac{3}{2}x - 3$ $\quad y = \frac{-3}{2}x + 3$

(line 1)   (line 2)

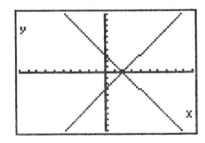

Solution is (2,0)

Check: Substitue 2 for x and 0 for y in both equations.

$3x - 2y = 6$ $3x + 2y = 6$
$3(2) - 2(0) = 6$ $3(2) + 2(0) = 6$
$6 - 0 = 6$ $6 + 0 = 6$
$\quad\quad 6 = 6$ True $\quad\quad 6 = 6$ True

b. $3x - y = 9$
   $6x - 2y = 12$

Solution: Solve each equation for y and graph using the slope and y-intercept.

$3x - y = 9$
$ -y = -3x + 9$
$y = 3x - 9$
   (line 1)

$6x - 2y = 12$
$ -2y = -6x + 12$
$y = 3x - 6$
   (line 2)

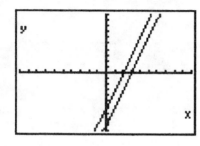

From the graph the lines appear to be parallel. The system is inconsistent and has no solution.

Exercise set 8.1

Determine which, if any, of the following ordered pairs satisfy each system of linear equations.

1. $x + y =$
   $x - y =$
   (a)   (b) (2,1)   (c) 8,1)

2. $x + y =$
   $2x + y = 12$
   (a) (6,2)   (b) (4,4)   (c) (3,6)

3. $x + y = 3$
   $x - 2y = 12$
   (a) (1,2)   (b) (0,-6)   (c) (6,-3)

4. $2x + 4y = 9$
   $x + y = 1$
   (a) (4,-3)   (b) (0, 9/4)   (c) (8,-7)

Express each equation in y-intercept form. Without graphing the equations, state whether the system has one solution, no solution, or an infinite number of solutions.

5. $4x - y = 6$
   $2x + y = 3$

6. $x + y = 8$
   $4x + 4y = 10$

7. 7x - 3y = 11
   2x + 5y = 7

8. 2x - 4y = 12
   -x + 2y = -6

Determine the solution to the following systems of equations graphically. If the system is dependent or inconsistent, so state.

9. 7x - y = 7
   x - y = 1

10. 2x + y = 8
    2x - y = -4

11. x + y = 5
    2x + 2y = 12

12. x - y = 2
    3x - 3x = -6

13. x = 2y + 5
    x = y + 1

14. 2x - 3y = -11
    x + 3y = 8

Answers to exercise set 8.1

1. (2,-5)   2. (4,4)   3. (6,-3)   4. none

5. one solution   6. no solution   7. one solution

8. infinite number of solutions   9. (1,0)   10. (1,6)

11. inconsistent   12. dependent   13. (-3,-4)

14. (-1,3)

8.2 Solving Systems of Equations by Substitution

Summary

> **To Solve a System of Equations by Substitution**
>
> 1. Solve for a variable in either equation. (If possible, solve for a variable with a numerical coefficient of 1 to avoid working with fractions.)
>
> 2. Substitute the expression found for the variable in step 1 into the other equation.
>
> 3. Solve the equation determined in step 2 to find the value of one variable.
>
> 4. Substitute the value of the variable determined in step 3 into the equation obtained in step 1 to find the other variable.

Example 1. Solve the following system of equations by substitution.

$$x + y = 3$$
$$2x - 3y = 1$$

1. Solve for y in the first equation.
$$x + y = 3$$
$$y = 3 - x$$

2. Substitute $3 - x$ for y in the second equation.
$$2x - 3(3 - x) = 1$$
3.
$$2x - 9 + 3x = 1$$
$$5x - 9 = 1$$
$$5x = 10$$
$$x = 2$$

4. Now, substitute $x = 2$ in the equation $y = 3 - x$
$$y = 3 - x$$
$$y = 3 - 2$$
$$y = 1$$

So, the solution is the ordered pair (2,1)

Example 2. Solve the following system of equations by substitution.

$$2x + 3y = 3$$
$$6x - 3y = 1$$

1. Solve for x in the first equation. (Note, since there are no variables with a coefficient of 1, we cannot avoid fractions.)

$$2x + 3y = 3$$
$$2x = 3 - 3y$$
$$x = \frac{3 - 3y}{2}$$
$$x = \frac{3}{2} - \frac{3y}{2}$$

2. Substitute $\frac{3}{2} - \frac{3y}{2}$ for x in the second equation.

$$6\left(\frac{3}{2} - \frac{3y}{2}\right) - 3y = 1$$

3.
$$9 - 9y - 3y = 1$$
$$9 - 12y = 1$$
$$-12y = -8$$
$$y = \frac{-8}{-12} = \frac{2}{3}$$

4. Now, substitute y = 2/3 in the equation $x = \frac{3}{2} - \frac{3y}{2}$

$$x = \frac{3}{2} - \frac{3y}{2}$$
$$x = \frac{3}{2} - \frac{3}{2}\left(\frac{2}{3}\right)$$
$$x = \frac{3}{2} - 1$$
$$x = \frac{1}{2}$$

So, the solution is (1/2, 2/3).

Example 3. Solve the following system of equations by substitution.

$$y = 2x + 5$$
$$1 = 6x - 3y$$

1. The first equation is already solved for y. So, we will use $y = 2x + 5$.

2. Substitute $2x + 5$ for y in the second equation.
   $$6x - 3(2x + 5) = 1$$

3. $6x - 6x - 15 = 1$
   $$-15 = 1$$

Since the statement $-15 = 1$ is false, the system has no solution. The graphs of the equations will be parallel lines and the system is inconsistent.

Example 4. Solve the following system of equations by substitution.

$$x - 2y = 3$$
$$3x - 6y = 9$$

1. Solve for x in the **first** equation.

   $$x - 2y = 3$$
   $$x = 3 + 2y$$

2. Substitute $3 + 2y$ for x in the second equation.

   $$3(3 + 2y) - 6y = 9$$
   $$9 + 6y - 6y = 9$$
   $$9 = 9$$

Since $9 = 9$ is a true statement, this system has an infinite number of solutions. The graphs of the equations represent the same line when graphed and the system is dependent.

Exercise set 8.2

Find the solution to each system of equations

1. $y = x - 5$
   $x + y = 3$

2. $x = y + 1$
   $3x + 4y = 17$

3. $x + y = 9$
   $8x - y = -18$

4. $x + 8y = 6$
   $-2x - 4y = 3$

5. $5x = y + 8$

6. $2x + 3y = 8$

$$10x - 2y = 16 \qquad\qquad y = \frac{-3}{2}x - 2$$

7. $2x + y = 3$
    $4x + 3y = 10$

8. $x = 4y + 1$
    $3x - 8y = 8$

9. $2x + 2y = -12$
    $5y = 3x - 6$

10. $4x - 3y = 1$
    $5x + 2y = 7$

Answers to exercise set 8.2

1. (4,-1)   2. (3,2)   3. (-1, 10)   4. $(-4, \frac{5}{4})$

5. infinite number of solutions   6. no solution   7. $(\frac{-1}{2}, 4)$

8. $(6, \frac{5}{4})$ 9. (-3, -3)   10. (1, 1)

8.3 Solving systems of Equations by the Addition Method

Summary

> **To Solve a System of Equations by the Addition (or Elimination) Method**
>
> 1. If necessary, rewrite each equation so that the terms containing the variables appear on the left hand side of the equal sign and any constants appear on the right side of the equal sign.
>
> 2. If necessary, multiply both equations by a constant(s) so that when the equations are added the resulting sum will contain only one variable.
>
> 3. Add the equations. This will result in a single equation containing only one variable.
>
> 4. Solve for the variable in the equation in step 3.
>
> 5. Substitute the value found in step 4 into either of the original equations. Solve that equation to find the value of the remaining variable.

Example 1 Solve the following equation using the addition method

$$3x - 2y = 10$$
$$3x + 2y = 14$$

Solution: Steps 1 and 2 are unnecessary.

Step 3.
$$\begin{aligned} 3x - 2y &= 10 \\ \underline{3x + 2y} &= \underline{14} \\ 6x &= 24 \end{aligned}$$

Step 4. $6x = 24$

$x = \dfrac{24}{6}$

$x = 4$

Step 5. Substitute $x = 4$ into the second equation
$$3x + 2y = 14$$
$$3(4) + 2y = 14$$
$$12 + 2y = 14$$
$$2y = 2$$
$$y = \dfrac{2}{2} = 1$$

So, the solution to the system of equations is (4,1)

Example 2. Solve the following system of equations using the addition method

$$2x + 3y = 1$$
$$3x = 2y + 16$$

Step 1. $2x + 3y = 1$
$\underline{3x - 2y = 16}$

Step 2. Multiply the first equation by 2 and the second equation by 3 so that upon addition the y variable will be eliminated.

$2[\,2x + 3y = 2\,]$ gives $\quad 4x + 6y = 4$
$3[\,3x - 2y = 16\,]$ gives $\quad 9x - 6y = 48$

Step 3. $4x + 6y = 4$
$\underline{9x - 6y = 48}$
$13x = 52$

Step 4. $13x = 52$

$x = \dfrac{52}{13} = 4$

Step 5. Substitute $x = 4$ into the first equation.

$$2x + 3y = 2$$
$$2(4) + 3y = 2$$
$$8 + 3y = 2$$
$$3y = -6$$

$$y = \frac{-6}{3} = -2$$

So, the solution to the system of equations is (4,-2)

Example 3. Solve the following system of equations using the addition method.

$$3x - 6y = 8$$
$$x - 2y = 2$$

Solution: Step 1 is unnecessary.

Step 2. Multiply the second equation by -3 so that upon addition the y variable will be eliminated.

$$3x - 6y = 8$$ $$3x - 6y = 8$$
$$-3[x - 2y = 2] \text{ gives}$$ $$-3x + 6y = -6$$

Step 3. $3x - 6y = 8$
         $\underline{-3x + 6y = -6}$
         $0 = 2$

Step 4. Since 0 = 2 is a false statement, the system has no solution. The graphs of the equations are parallel lines and the system is inconsistent.

Example 4. Solve the following system of equations using the addition method.

$$4x + 8y = 12$$
$$x = 3 - 2y$$

Solution:

Step 1. $4x + 8y = 12$
          $x + 2y = 3$

Step 2.   Multiply the second equaion by -4 to eliminate the x variable.

$$4x + 8y = 12$$ $$4x + 8y = 12$$
$$-4[x + 2y = 3] \text{ gives}$$ $$-4x - 8y = -12$$

Step 3. $4x + 8y = 12$
          $\underline{-4x - 8y = -12}$
          $0 = 0$

Step 4 Since 0 = 0 is a true statement, there is an infinite number of solutions to the system. The graphs of the equations will be the same line and the system is dependent.

Exercise set 8.3

Solve the following systems of equations using the addition method.

1. $x + y = -1$
    $x - y = 3$

2. $2x - y = 3$
    $-2x + y = 2$

3. $x + 5y = 4$
    $-x - 5y = -4$

4. $3x - 2y = -11$
    $x = -1 - 2y$

5. $y = 1 - 2x$
    $6x + 5y = 13$

6. $x + 4y = 2$
    $x - 8y = -7$

7. $2x + 4y = 8$
    $3x + 6y = 6$

8. $7x - 5y = 19$
    $3x - 2y = 6$

9. $3x - 5y = 10$
    $7x - 3y = 14$

10. $5x + 7y = 1$
    $7x + 5y = -1$

Answers to Exercise set 8.3

1. (1,-2)   2. no solution   3. infinite number of solutions

4. (-3,1)   5. (-2,5)   6. $(-1, \frac{3}{4})$

7. no solution   8. (-8, -15)   9. $\left(\frac{20}{13}, \frac{-14}{13}\right)$

10. $\left(\frac{-1}{2}, \frac{1}{2}\right)$

8.4 Applications of Systems of Equations

Summary

> Many of the application problems solved in earlier chapters using only one variable can also be solved using two variables.

Example 1 Two angles are supplementary when the sum of their measures is 180°. Angles x and y are supplementary and angle x is 26° more than angle y. Find angles x and y.

Solution: Let x = measure of angle x
Let y = measure of angle y

$$x = 26 + y$$
$$x + y = 180 \text{ (since x and y are supplementary)}$$

To solve the system, use the addition method.

$$\begin{aligned} x - y &= 26 \\ \underline{x + y} &= \underline{180} \\ 2x &= 206 \\ x &= 103 \end{aligned}$$

Substitute x = 103 in the second equation.

$$\begin{aligned} x + y &= 180 \\ 103 + y &= 180 \\ y &= 77 \end{aligned}$$

So, angle x has measure of 103° and angle y has measure 77°.

Example 2 A box contains quarters and nickels. The number of nickels is four times the number of quarters. The total worth of the coins is $1.35. Find the number of nickels and the number of quarters in the box.

Solution: Let x = number of nickels
Let y = number of quarters

$$x = 4y \text{ (the number of nickels is 4 times the number of quarters)}$$

$$5x + 25y = 135 \text{ (the worth of the coins is \$1.35)}$$

Use substitution to solve this system. Substitute x = 4y into the second equation

$$5(4y) + 25y = 135$$
$$20y + 25y = 135$$
$$45y = 135$$
$$y = \frac{135}{45} = 3$$

Now substitute 3 for y in the first equation.
$$x = 4y$$
$$x = 4(3) = 12$$

So there are 12 nickels and 3 quarters in the box.

Example 3. A boat can make a 60 mile trip upstream (against the current) in 5 hours. The return trip downstream (with the current) takes 3 hours. Find the speed of the boat in still water and the speed of the current.

Solution: Let x = speed of the boat in still water
Let y = speed of the current

Boat's direction	Rate	Time	Distance
upstream	x - y	5	60
downstream	x + y	3	60

So the system of equations is
$$5(x - y) = 60$$
$$3(x + y) = 60$$

Use addition method to solve.

$$3[\,5x - 5y = 60\,]$$
$$5[\,3x + 3y = 60\,]$$

gives
$$15x - 15y = 180$$
$$15x + 15y = 300$$
$$30x = 480$$
$$x = 16$$

Now, substitute x = 16 into the second equation.

$$3x + 3y = 60$$
$$3(16) + 3y = 60$$
$$48 + 3y = 60$$
$$3y = 12$$
$$y = \frac{12}{3}$$
$$y = 4$$

So the speed of the boat in still water is 16 miles per hour and the speed of the current is 4 miles per hour.

Example 4. A total of $80,000 is to be invested in two accounts. One account earns 10% simple interest and the other earns 6% simple interest. Find the amount invested in each account if the total interest to be earned after one year is $6,000.

Solution: Let x = amount invested at 10%
             Let y = amount invested at 6 %

	Principal	Rate	Time	Interest
10% account	x	.10	1	.10x
6% account	y	.06	1	.06y

The system of equations is $.10x + .06y = 6,000$
                              $x + y = 80,000$

We will use the substitution method to solve the system.

Since $x + y = 80,000$, $y = 80,000 - x$.

Substitute $80,000 - x$ for y in the first equation.

$.10x + .06(80,000 - x) = 6,000$
$.10x + 4,800 - .06x = 6,000$
$.04x + 4,800 = 6,000$
$.04x = 1,200$

$$x = \frac{1200}{.04}$$

$x = 30,000$

Now, substitute $x = 30,000$ into the second equation.

$x + y = 80,000$
$30,000 + y = 80,000$
$y = 50,000$

So, $30,000 should be invested at 10% and $50,000 should be invested at 6% .

Exercise set 8.4

Express each exercise as a system of linear equations, and then find the solution. Use a calculator where appropriate.

1. Two angles are complementary. If the measure of one angle is twice that of the second angle, find the measures of the two

angles.

2. Two angles are supplementary. If the measure of one angle is 30 degrees more than twice the second angle, find the measure of the two angles.

3. A box contains dimes and quarters. The total value of the coins in the box is $3.00. If there are 5 more quarters than dimes, find the number of dimes and the number of quarters.

4. A piggy bank contains pennies and nickels. The total value of the coins in the piggy bank is 80 cents. If the number of pennies is 10 more than twice the number of nickels, find the number of pennies and the number of nickels.

5. A rectangular lot has a perimeter of 200 feet. If the length of the lot is 20 feet more than the width of the lot, determine the dimensions.

6. A rectangle has a perimeter of 48 inches. If the length of the rectangle is twice the width, determine the rectangle's dimensions.

7. Paddling with the current, a canoist can go 24 miles in 3 hours. Against the current it takes 4 hours to go the same distance. Find the rate of the canoist in still water and the rate of the current.

8. A rowing team rowing with the current traveled 16 miles in 2 hours. Against the current, the team rowed 8 miles in 2 hours. Find the rate of the rowing team in still water and the rate of the current.

9. A total of $20,000 is to be invested in two accounts. The first account earns 5% simple interest whereas the second account earns 9% simple interest. Find the amount invested in each account if the total interest to be earned after one year is $1200.

10. A landscaper wants 600 pounds of grass seed that is 45% bluegrass. She has two mixtures available. One contains 40% bluegrass and the other 70% bluegrass. How many pounds of each must she use ?

Answers to exercise set 8.4

1. $x + y = 90$
   $x = 2y$
   one angle = $30°$, other angle $60°$

2. $x + y = 180$
   $y = 2x + 30$
   one angle = $50°$
   other angle = $130°$

3. $y = x + 5$
   $25y + 10x = 300$

4. $y = 2x + 10$
   $y + 5x = 80$

       5 dimes; 10 quarters   10 nickels; 30 pennies

5. $2x + 2y = 200$ 6. $2x + 2y = 48$
    $y = x + 20$ $y = 2x$
   length = 60 ft.; width = 40 ft.   l = 16 in.; w = 8 in.

7. $(x + y)(3) = 24$
   $(x - y)(4) = 24$
   rate of boat = 7 mph
   rate of current = 1 mph

8. $(x + y)(2) = 16$
   $(x - y)(2) = 8$
   rate of rowing team = 6 mph
   rate of current = 2 mph

9. $x + y = 20000$
   $.05x + .09y = 1200$
   $15,000 at 5%; $5,000 at 9%

10. $x + y = 600$
    $.40x + .70y = 270$
    500 lb. of 40% mixture
    100 lb. of 70% mixture

8.5 Systems of Linear Inequalities
Summary

> **To Solve a System of Linear Inequalities**
>
> Graph each inequality on the same set of axes. The solution set is the set of points that satisfies all the inequalities in the system.

Example 1. Determine the solution to the system of inequalities.

$$y \leq 2x - 1$$
$$x + 2y > 4$$

Solution. First graph $y \leq 2x - 1$. To graph this inequality, first graph the line $y = 2x - 1$. This line has a slope of 2 and y intercept of -1 . The line is solid since the points on the line are included due to the inequality symbol $\leq$. Now, pick a test point not on the line. Usually, the origin is a good choice. If (0,0) is substituted into the inequality $y \leq 2x - 1$, the inequality $0 \leq 2(0) - 1$ is obtained. The inequality is false. Therefore, the region **to the right of** the line is shaded. Using the above procedure, graph the inequality $x + 2y > 4$ on the same set of axes. The solution is the part of the two graphs where the two shaded regions intersect.

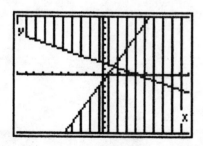

Example 2. Determine the solution to the system of inequalities

$$x > 2$$
$$y < x - 2$$

Solution. First graph $x > 2$. Graph $x = 2$ first. This graph represents a vertical line through $(2,0)$. The line is dotted because of the inequality symbol, $>$. Now, shade the region **to the right of** the vertical line $x = 2$. Now, graph the line $y = x - 2$. The line has a slope of 1 and y intercept of -2. This line is also dotted. Substitute the origin, $(0,0)$, in the inequality $y < x - 2$ to obtain $0 < 0 - 2$, which is false. Therefore, the region below the line $y = x - 2$ which **does not include the origin** is shaded. The solution to the system is the region of intersection.

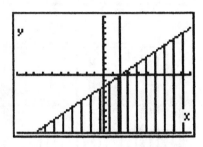

Exercise set 8.5

Determine the solution to each system of inequalities.

1. $x - y < 4$
   $y \geq x + 2$

2. $y \geq \frac{1}{3}x$
   $y > -x + 1$

3. $y > x$
   $y \geq 4$

4. $y < 1 - x$
   $x - 2y \leq 4$

5. $y \leq x$
   $x + 2y \leq 1$

6. $x < 3y - 3$
   $x \geq \frac{1}{2}y$

Answers to set 8.5

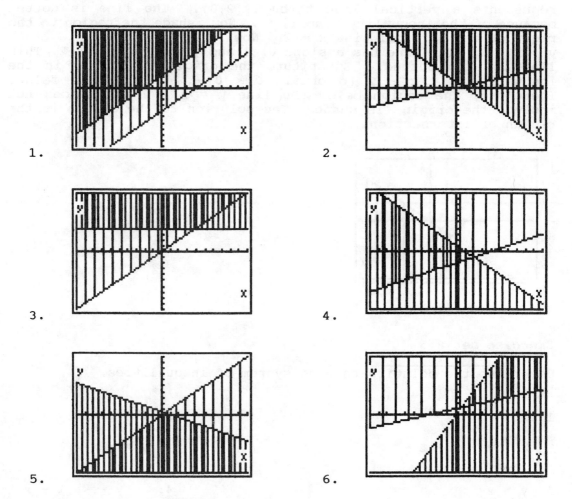

1.

2.

3.

4.

5.

6.

Practice Test Chapter 8

1. Determine which, if any of the ordered pairs satisfy the system of equations

$$3x + 2y = 10$$
$$x - y = 0$$

a) (5,5)   b) (2,2)   c) (0,5)

Identify each system as consistent, inconsistent, or dependent. State whether the system has exactly one solution, no solution, or an infinite number of solutions.

2.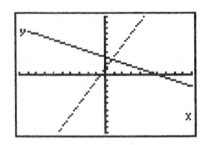

3.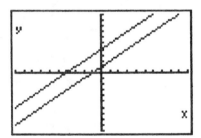

Write each equation in slope-intercept form. Then determine, without solving the system, whether the system of equations has exactly one solution, no solution, or an infinite number of solutions.

4. $2x - 3y = 6$
   $-4x + 6y = -12$

5. $3x + 2y = 7$
   $2x - 3y = -3$

Solve the following systems of equations graphically.

6. $y = 2x + 1$
   $y = x - 2$

7. $2x - y = 3$
   $x - y = -3$

Solve the following systems of equations using substitution.

8. $y = 2x - 1$
    $2x + y = 11$

9. $9x - 2y = 2$
    $3x + 8y = 70$

Solve the following systems of equations using the addition method.

10. $2x - 3y = 11$
    $5x + y = 19$

11. $5x + 11y = 19$
    $2x - 3y = -22$

Express the problem as a system of linear equations, and then find the solution.

12. A boat can go downstream (with the current) a distance of 45 miles in 5 hours. The return trip (against the current) takes 9 hours. Find he rate of teh boat in still water and the rate of the current.

13. A box contains nickels and dimes. The total value of the coins is $1.40. If the number of dimes is 4 more than twice the number of nickels, find the number of nickels and the number of dimes.

14. Determine the solution to the system of inequalities.

    $x + y < 5$
    $x - 3y \geq 9$

Answers to Chapter 8 Practice Test

1. (2,2)
2. consistent, one solution
3. inconsistent no solution

4. infinite number of solutions
5. one solution

6. (-3, -5)
7. (6,9)
8. (3,5)
9. (2,8)

10. (4,-1)
11. (-5,4)
12. $(x + y)(5) = 45$
    $(x - y)(9) = 45$
    rate of boat = 7 mph
    rate of current = 2 mph

13. $5x + 10y = 140$
    $y = 2x + 4$

    12 dimes; 4 nickels

14.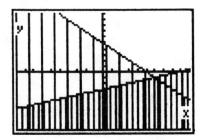

9.1 Evaluating Square Roots

Summary:

> 1. $\sqrt{}$ is called a radical sign.
>
> 2. The expression underneath the radical sign is called the **radicand**.
>
> 3. The **entire** expression is called a **radical expression**.
>
> 4. The **index** tells the "root" of the expression.
>
> 5. The **principal** or **positive** square root of a positive number is denoted by $\sqrt{x}$ and is that **positive number** whose square is x.
>
> 6. Square roots of negative numbers are not real numbers. The square root of a negative number is an imaginary number.

Example 1

$\sqrt{27}$ is read as "the square root of 27". The radicand is 27. The index is understood to be 2. $\sqrt{27}$ represents that positive number whose square is 27.

Example 2

$\sqrt[3]{64}$ is read as "the cube root of 64". The radicand is 64. The index is 3. Cube roots will be discussed at a later time.

Example 3

Evaluate the following:

a) $\sqrt{81}$ b) $\sqrt{121}$ c) $-\sqrt{49}$ d) $-\sqrt{36}$

Example 3 continued:

a) $\sqrt{81}$ is 9 since $9^2 = 81$. b) $\sqrt{121}$ is 11 since $11^2 = 121$.

c) $-\sqrt{49}$ is the opposite of $\sqrt{49}$. $\sqrt{49}$ is 7 since 7^2 is 49 so $-\sqrt{49}$ is -7.

d) $-\sqrt{36}$ is the opposite of $\sqrt{36}$ or -6.

Example 4. Indicate whether the radical expression is a real or imaginary number.

a) $-\sqrt{16}$ b) $\sqrt{-16}$ c) $\sqrt{-54}$ d) $-\sqrt{54}$

a) $-\sqrt{16}$ is the opposite of $\sqrt{16}$ or -4 which is a real number.

b) $\sqrt{-16}$ is an imaginary number since the radicand is negative.

c) $\sqrt{-54}$ is an imaginary number since the radicand is negative.

d) $-\sqrt{54}$ represents the opposite of $\sqrt{54}$ which is a real number.

Example 5. Use your calculator to evaluate each square root.

a) $\sqrt{12}$ b) $\sqrt{27}$ c) $\sqrt{13}$ d) $\sqrt{123}$

a) about 3.46   b) about 5.196   c) about 3.606
d) about 11.09

Summary

> 1. When a perfect square number appears under the radican, its square root may be taken exactly.
>
> 2. The first ten perfect square numbers are 1, 4, 9, 16, 25, 36, 49, 64, 81, 100.

Example 6. Answer true or false.

a) $\sqrt{36}$ is a rational number.   b) $-\sqrt{4}$ is not a real number.

c) $\sqrt{89}$ is rational.   d) $\sqrt{\frac{9}{49}}$ is a rational number.

solution: a) True because 36 is a perfect square.

b) False. $-\sqrt{4}$ is the opposite of $\sqrt{4}$ or -2. -2 is a real number.

c) False. 89 is not a perfect square number. When its square root is taken, the result will be an irrational number. **Any** number whose root cannot be taken **exactly** is an irrational number.

d) True. $\sqrt{\frac{9}{49}}$ is 3/7 since $(3/7)^2 = 9/49$.

Summary

> 1. To write a square root in exponential form, the radicand is raised to the exponent 1/2.

Example 7.   Square root form: Exponential form:

$$\sqrt{64}$$
$$(64)^{1/2}$$

$$\sqrt{5x}$$
$$(5x)^{1/2}$$

Example 8.   Convert the expressions below to radical form

a) $(x^4)^{1/2} = \sqrt{x^4} = x^2$ since $(x^2)^2 = x^4$.

b) $(9x)^{1/2} = \sqrt{9x}$

Exercise set 9.1

Evaluate each square root

1. $\sqrt{144}$

2. $\sqrt{\dfrac{16}{25}}$

3. $\sqrt{\dfrac{100}{169}}$

4. $\sqrt{\dfrac{25}{64}}$

5. $-\sqrt{169}$

6. $-\sqrt{\dfrac{64}{81}}$

Use your calculator to evaluate each square root. Round all answers to three decimal places.

7. $\sqrt{89}$

8. $\sqrt{24}$

9. $-\sqrt{56}$

Answer true or false.

10. $\sqrt{49}$ is rational.

11. $-\sqrt{121}$ is not a real number.

12. $\sqrt{13^2}$ is an integer.

13. $\sqrt{113}$ is a rational number.

14. $\sqrt{\frac{121}{100}}$ is a rational number.

15. $\sqrt{-36}$ is not a real number.

Write in exponential form.

16. $\sqrt{12}$

17. $\sqrt{7x}$

18. $\sqrt{10x^2}$

19. $\sqrt{60x^3}$

20. $\sqrt{49x^3y^3}$

Answers to exercise set 9.1

1. 12   2. 4/5   3. $\frac{10}{13}$ 4. 5/8   5. -13   6. -8/9

7. 9.434   8. 4.899   9. -7.483   10. T   11. F
12. T   13. F   14. T   15. T   16. $(12)^{1/2}$
17. $(7x)^{1/2}$ 18. $(10x^2)^{1/2}$ 19. $(60x^3)^{1/2}$
20. $(49x^3y^3)^{1/2}$

9.2 Multiplying and Simplifying Square Roots

Summary:

1. $\sqrt{a} \sqrt{b} = \sqrt{ab}$, provided $a \geq 0$, and $b \geq 0$.

2. To simplify the square root of a constant:

   a) Write the constant as a product of the largest perfect square factor and another factor.

   b) Use the product rule to write the expression as a product of square roots, each square root containing one of the factors.

   c) Find the square root of the perfect square factor.

Example 1. Simplify $\sqrt{50}$

$$\sqrt{50} = \sqrt{25 \cdot 2} = \sqrt{25} \sqrt{2} = 5\sqrt{2}$$

Example 2. Simplify $\sqrt{128}$

$$\sqrt{128} = \sqrt{64 \cdot 2} = \sqrt{64} \sqrt{2} = 8\sqrt{2}$$

Example 3. Simplify $\sqrt{192}$

$$\sqrt{192} = \sqrt{64 \cdot 3} = \sqrt{64} \sqrt{3} = 8\sqrt{3}$$

Summary:

> 1. The square root of a variable raised to an even power equals the variable raised to one-half that power.
>
> 2. To simplify the square root of a radicand containing a variable raised to an odd power, do the following:
>
> a) express the variable as the product of two factors, one of which has an exponent of 1.
>
> b) Use the product rule to simplify.

In the examples which follow, assume that all variables represent non-negative numbers.

Simplify the following.

Example 4. a) $\sqrt{a^4} = a^{4/2} = a^2$, since $(a^2)^2 = a^4$

b) $\sqrt{x^{20}} = x^{20/2} = x^{10}$, since $(x^{10})^2 = x^{20}$.

c) $\sqrt{a^4 \cdot b^{10}} = \sqrt{a^4} \sqrt{b^{10}} = a^{\frac{4}{2}} \cdot b^{\frac{10}{2}} = a^2 \cdot b^5$

Example 5. a) $\sqrt{y^5} = \sqrt{y^4 \cdot y} = \sqrt{y^4} \cdot \sqrt{y} = y^2 \cdot y$

b) $\sqrt{a^{49}} = \sqrt{a^{48} \cdot a} = \sqrt{a^{48}} \cdot \sqrt{a} = a^{\frac{48}{2}} \sqrt{a} = a^{24} \sqrt{a}$

Example 6. a) $\sqrt{98x^3} = \sqrt{49x^2 \cdot 2x} = \sqrt{49x^2} \cdot \sqrt{2x} = 7x\sqrt{2x}$

b) $\sqrt{128x^9y^7} = \sqrt{64x^8y^6 \cdot 2xy}$

$= \sqrt{64x^8y^6} \cdot \sqrt{2xy}$

$= 8x^4y^3 \ \sqrt{2xy}$

Example 7. Multiply and then simplify.

a) $\sqrt{6}\sqrt{3} = \sqrt{6 \cdot 3} = \sqrt{18} = \sqrt{9 \cdot 2} = \sqrt{9}\sqrt{2} = 3\sqrt{2}$

b) $\sqrt{27} \cdot \sqrt{3} = \sqrt{27 \cdot 3} = \sqrt{81} = 9$

c) $\sqrt{\frac{1}{2}x \cdot \sqrt{18x}} = \sqrt{\frac{1}{2}x \cdot 18x} = \sqrt{9x^2} = \sqrt{9}\sqrt{x^2} = 3x$

d) $\sqrt{20xy^4} \cdot \sqrt{6x^5} =$

$\sqrt{20xy^4 \cdot 6x^5} =$

$\sqrt{120x^6y^4} =$

$\sqrt{4 \cdot x^6y^4 \cdot 30} =$

$\sqrt{4x^6y^4} \cdot \sqrt{30} = 2x^3y^2\sqrt{30}$

Exercise set 9.2

Simplify each expression.

1. $\sqrt{75}$ 5. $\sqrt{108}$ 9. $\sqrt{x^3y^3z^3}$ 13. $\sqrt{40} \cdot \sqrt{5}$

2. $\sqrt{100}$ 6. $-\sqrt{121}$ 10. $\sqrt{64xyz^4}$ 14. $\sqrt{3b}\sqrt{12b^4}$

3. $\sqrt{44}$ 7. $\sqrt{a^8}$ 11. $\sqrt{8} \cdot \sqrt{12}$ 15. $\sqrt{4x}\sqrt{xy}$

4. $\sqrt{54}$ 8. $\sqrt{x^4 y}$ 12. $\sqrt{5} \cdot \sqrt{10}$

16. $(\sqrt{3x})^2 (\sqrt{9x})^2$

17. $\sqrt{15x^2} \sqrt{6x^5}$ 18. $\sqrt{18x^4} \sqrt{10x^2}$ 19. $\sqrt{x^{101}}$

20. $-\sqrt{500}$

Answers to exercise set 9.2

1. $5\sqrt{3}$ 2. 10   3. $2\sqrt{11}$ 4. $3\sqrt{6}$ 5. $6\sqrt{3}$

6. -11 7. a^4 8. $x^2 \sqrt{y}$ 9. $xyz \sqrt{xyz}$ 10. $8z^2 \sqrt{xy}$

11. $4\sqrt{6}$ 12. $5\sqrt{2}$ 13. $10\sqrt{2}$ 14. $6b^2 \sqrt{5}$ 15. $2x\sqrt{y}$

16. $27x^2$ 17. $3x^3 \sqrt{10x}$ 18. $6x^3 \sqrt{5}$ 19. $x^{50} \sqrt{x}$

20. $-10\sqrt{5}$

9.3 Dividing and Simplifying Square Roots

Summary:

> A square root is simplified when:
>
> 1. **No** radicand has a factor which is a perfect square.
>
> 2. **No** radicand contains a fraction.
>
> 3. **No** denominator contains a square root.

Example 1. Explain why the following radical expressions **are not** simplified.

a) $\sqrt{20}$ — Since 20 = 4(5) and 4 is a perfect square.
The radicand contains a factor which is a **perfect** square.

b) $\sqrt{\dfrac{1}{3}}$ — Radicand contains a fraction.

c) $\dfrac{1}{\sqrt{5}}$ — Denominator contains a square root.

Summary:

> **Quotient Rule for Radicals**
>
> $\dfrac{\sqrt{a}}{\sqrt{b}} = \sqrt{\dfrac{a}{b}} \qquad a \geq 0, \; b \geq 0$

Example 2. Simplify.

a) $\sqrt{\dfrac{50}{2}}$ b) $\sqrt{\dfrac{54}{2}}$ c) $\sqrt{\dfrac{16x^3}{4x}}$ d) $\sqrt{\dfrac{64}{49}}$

Solution:

a) $\sqrt{\dfrac{50}{2}} = \sqrt{25} = 5$

b) $\sqrt{\dfrac{54}{2}} = \sqrt{27} = \sqrt{9 \cdot 3} = \sqrt{9} \cdot \sqrt{3} = 3\sqrt{3}$

c) $\sqrt{\dfrac{16x^3}{4x}} = \sqrt{\dfrac{16}{4}x^2} = \sqrt{4x^2} = \sqrt{4}\sqrt{x^2} = 2x$

d) $\sqrt{\dfrac{64}{49}} = \dfrac{\sqrt{64}}{\sqrt{49}} = \dfrac{8}{7}$

Example 3. Simplify.

a) $\dfrac{\sqrt{108}}{\sqrt{3}}$ b) $\dfrac{\sqrt{12x^5}}{\sqrt{3x}}$ c) $\dfrac{\sqrt{75a^3b^2}}{\sqrt{3ab}}$

Solution:

a) $\dfrac{\sqrt{108}}{\sqrt{3}} = \sqrt{\dfrac{108}{3}} = \sqrt{36} = 6$

b) $\dfrac{\sqrt{12x^5}}{\sqrt{3x}} = \sqrt{\dfrac{12x^5}{3x}} = \sqrt{4x^4} = \sqrt{4}\sqrt{x^4} = 2x^2$

c) $\dfrac{\sqrt{75a^3b^2}}{\sqrt{3ab}} = \sqrt{\dfrac{75a^3b^2}{3ab}} = \sqrt{25a^2b} = \sqrt{25}\sqrt{a^2}\sqrt{b} = 5a\sqrt{b}$

Summary:

> 1. **To rationalize a denominator** means to remove all radicals from the denominator.
>
> 2. **To rationalize a denominator**, multiply both the numerator and the denominator of the fraction by the square root that appears in the denominator or by the square root of a number that makes the denominator a perfect square.

Example 4. Simplify.

a) $\dfrac{3}{\sqrt{3}}$ b) $\dfrac{6}{\sqrt{5}}$ c) $\sqrt{\dfrac{3}{7}}$ d) $\sqrt{\dfrac{y^2}{50}}$

Solution.

a) $\dfrac{3}{\sqrt{3}} = \dfrac{3\sqrt{3}}{\sqrt{3}\sqrt{3}} = 3\dfrac{\sqrt{3}}{\sqrt{9}} = \dfrac{3\sqrt{3}}{3} = \sqrt{3}$

b) $\dfrac{6}{\sqrt{5}} = \dfrac{6\sqrt{5}}{\sqrt{5}\sqrt{5}} = \dfrac{6\sqrt{5}}{\sqrt{25}} = 6\dfrac{\sqrt{5}}{5}$

c) $\sqrt{\dfrac{3}{7}} = \dfrac{\sqrt{3}}{\sqrt{7}} = \dfrac{\sqrt{3}\sqrt{7}}{\sqrt{7}\sqrt{7}} = \dfrac{\sqrt{3\cdot 7}}{\sqrt{49}} = \dfrac{\sqrt{21}}{7}$

d) $\sqrt{\dfrac{y^2}{50}} = \dfrac{\sqrt{y^2}}{\sqrt{50}} = \dfrac{y}{\sqrt{25\cdot 2}} = \dfrac{y}{\sqrt{25}\cdot\sqrt{2}} = \dfrac{y}{5\sqrt{2}}$

$= \dfrac{y\sqrt{2}}{5\sqrt{2}\sqrt{2}} = \dfrac{\sqrt{2}y}{5(2)} = \dfrac{\sqrt{2}y}{10}$

Exercise set 9.3

Simplify each expression.

1. $\sqrt{\dfrac{98}{2}}$ 2. $\sqrt{\dfrac{100}{5}}$ 3. $\sqrt{\dfrac{36}{9}}$ 4. $\dfrac{\sqrt{6}}{\sqrt{54}}$ 5. $\sqrt{\dfrac{10}{490}}$

6. $\sqrt{\dfrac{25}{64}}$ 7. $\dfrac{\sqrt{2}}{\sqrt{18}}$ 8. $\dfrac{\sqrt{72x^6 y^{10}}}{\sqrt{2x^2 y^2}}$ 9. $\dfrac{\sqrt{16y^4}}{\sqrt{8y}}$

10. $\dfrac{\sqrt{128x^{30}y^{40}}}{\sqrt{2x^{10}y^{20}}}$ 11. $\dfrac{2}{\sqrt{5}}$ 12. $\dfrac{7}{\sqrt{7}}$ 13. $\dfrac{5}{\sqrt{11}}$

14. $\dfrac{8}{\sqrt{50}}$ 15. $\dfrac{7}{\sqrt{72}}$ 16. $\sqrt{\dfrac{3}{5}}$ 17. $\sqrt{\dfrac{a^3}{18}}$

18. $\sqrt{\dfrac{a^2 b^2}{40}}$ 19. $\sqrt{\dfrac{75x^2}{3y}}$ 20. $\sqrt{\dfrac{40x^2 y^5}{6x^3 y^7}}$

Answers to exercise set 9.3

1. 7   2. $2\sqrt{5}$ 3. 2   4. 1/3   5. 1/7

6. 5/8   7. 1/3   8. $6x^2 y^4$ 9. $y\sqrt{2y}$ 10. $8x^{10}y^{10}$

11. $\dfrac{2\sqrt{5}}{5}$ 12. $\dfrac{2\sqrt{7}}{7}$ 13. $\dfrac{5\sqrt{11}}{11}$ 14. $\dfrac{4\sqrt{2}}{5}$

15. $\dfrac{7\sqrt{2}}{12}$ 16. $\dfrac{\sqrt{15}}{5}$ 17. $\dfrac{a\sqrt{2a}}{6}$ 18. $\dfrac{ab\sqrt{10}}{20}$

19. $\dfrac{5x\sqrt{y}}{y}$ 20. $\dfrac{2\sqrt{15x}}{3xy}$

9.4 Adding and Subtracting Square Roots

Summary:

> 1. **Like square roots** are square roots having the same radicands.
>
> 2. **To add like square roots**, add their coefficients and then multiply that sum by the like square root.

Simplify the following:

Example 1. $4\sqrt{5}+3\sqrt{5} = (4+3)\sqrt{5} = 7\sqrt{5}$

Example 2. $5\sqrt{a}+3\sqrt{a} = (5+3)\sqrt{a} = 8\sqrt{a}$

Example 3. $\dfrac{\sqrt{7}}{4}+3\dfrac{\sqrt{7}}{4} = \left(\dfrac{1}{4}+\dfrac{3}{4}\right)\sqrt{7} = \left(\dfrac{4}{4}\right)\sqrt{7} = \sqrt{7}$

Example 4. $5\sqrt{y} - 3\sqrt{y} + 4\sqrt{y} = (5-3+4)\sqrt{y} = 6\sqrt{y}$

Example 5. $\sqrt{x^2y} + 3\sqrt{x^2y} - \sqrt{x^2y} = (1+3-1)\sqrt{x^2y} = 3\sqrt{x^2y}$

Example 6. $\sqrt{3}+\sqrt{48} = \sqrt{3}+\sqrt{16}\sqrt{3} = 1\sqrt{3}+4\sqrt{3} = 5\sqrt{3}$

Example 7. $\sqrt{x^2y} + \sqrt{xy^2}+3\sqrt{xy^2}$

$\qquad = \sqrt{x^2y} + 4\sqrt{xy^2}$ since only the second and third radicals are like radicals

Example 8. $3\sqrt{18} - 5\sqrt{32} =$

$3\sqrt{9\cdot 2} - 5\sqrt{16\cdot 2} =$

$3\sqrt{9}\sqrt{2} - 5\sqrt{16}\sqrt{2} =$

$3(3)\sqrt{2} - 5(4)\sqrt{2} =$

$9\sqrt{2} - 20\sqrt{2} =$

$(9-20)\sqrt{2} =$

$-11\sqrt{2}$

Summary:

> 1. To **rationalize the denominator** of a rational expression, multiply both the numerator and denominator of the fraction by the **conjugate** of the denominator.
>
> 2. The **conjugate** of a binomial is a binomial having the same two terms with the sign of the second term changed

Example 9. Multiply $(3 + \sqrt{5})(3 - \sqrt{5})$ using the FOIL method.

$(3+\sqrt{5})(3-\sqrt{5})$
  F   O   I   L
$= 3(3) + 3(-\sqrt{5}) + 3\sqrt{5} - \sqrt{25}$
$= 9 - 5$
$= 4$

Example 10. Simplify $\dfrac{7}{3 + \sqrt{5}}$.

To rationalize the denominator, multiply both numerator and denominator by $3 - \sqrt{5}$, which is the conjugate of $3 + \sqrt{5}$.

$$\dfrac{7}{3 + \sqrt{5}} = \dfrac{7(3 - \sqrt{5})}{(3 + \sqrt{5})(3 - \sqrt{5})}$$

$$= \dfrac{7(3 - \sqrt{5})}{3^2 - (\sqrt{5})^2}$$

$$= \dfrac{7(3 - \sqrt{5})}{9 - 5} = \dfrac{7(3 - \sqrt{5})}{4}$$

Summary:

> To simplify a fraction whose numerator contains a constant term and a square root term:
>
> 1. Simplify the radical expression as far as possible.
>
> 2. If possible, factor out the GCF from each term in the numerator.
>
> 3. If possible, divide out a common factor from the GCF that was factored out in step 2 and the denominator.

Example 11. Simplify $\dfrac{2+\sqrt{12}}{6}$

$$\dfrac{2+\sqrt{12}}{6} = \dfrac{2+\sqrt{4\cdot 3}}{6}$$

$$= \dfrac{2+2\sqrt{3}}{6}$$

$$= \dfrac{2(1+\sqrt{3})}{6}$$

$$= \dfrac{1+\sqrt{3}}{3}$$

Exercise set 9.4

Simplify each radical expression.

1. $7\sqrt{10}-8\sqrt{10}$
2. $5\sqrt{3}+2\sqrt{3}-4\sqrt{3}$
3. $7\sqrt{2x}-3\sqrt{2x}$
4. $12\sqrt{15}+3\sqrt{15}-2\sqrt{15}$
5. $3+5\sqrt{y}-7\sqrt{y}$
6. $\sqrt{8}+5\sqrt{12}$
7. $\sqrt{27}+\sqrt{12}$
8. $\sqrt{48}+\sqrt{75}-\sqrt{27}$
9. $3\sqrt{17}+2\sqrt{19}$
10. $5\sqrt{108}-8\sqrt{180}$

Multiply as indicated

11. $(5+\sqrt{2})(5-\sqrt{2})$
12. $(\sqrt{7}+3)(\sqrt{7}-\sqrt{3})$
13. $(\sqrt{6}+\sqrt{3})(\sqrt{6}-\sqrt{3})$
14. $(\sqrt{x}+y)(\sqrt{x}-y)$
15. $(5\sqrt{2x}+\sqrt{5y})(5\sqrt{2x}-\sqrt{5y})$

Rationalize the denominator and simplify the radical expressions

16. $\dfrac{3}{\sqrt{7}+2}$ 17. $\dfrac{4}{\sqrt{x}+y}$ 18. $\dfrac{\sqrt{x}}{\sqrt{7}+\sqrt{x}}$

Simplify each radical expression if possible

19. $\dfrac{10+\sqrt{50}}{5}$ 20. $\dfrac{-50+\sqrt{200}}{40}$

Answers to exercise set 9.4

1. $-\sqrt{10}$ 2. $3\sqrt{3}$ 3. $4\sqrt{2x}$ 4. $13\sqrt{15}$ 5. $3-2\sqrt{y}$

6. $2\sqrt{2}+10\sqrt{3}$ 7. $5\sqrt{3}$ 8. $6\sqrt{3}$ 9. $3\sqrt{17}+2\sqrt{19}$

10. $30\sqrt{3}-48\sqrt{5}$ 11. 23 12. -2 13. 3

14. $x-y^2$ 15. $50x-5y$ 16. $\sqrt{7}-2$

17. $\dfrac{4(\sqrt{x}-y)}{x-y^2}$ 18. $\dfrac{\sqrt{x}(\sqrt{7}-\sqrt{x})}{7-x}$ 19. $2+\sqrt{2}$ 20. $\dfrac{-5+\sqrt{2}}{4}$

9.5 Solving Radical Equations

> **To Solve a Radical Equation Containing Only One Square Root Term**
>
> 1. Use the appropriate properties to rewrite the equation with the square root term by itself on one side of the equation. We call this **isolating** the radical.
>
> 2. Combine like terms.
>
> 3. Square both sides of the equation to remove the square root.
>
> 4. Solve the equation for the variable.
>
> 5. Check the solution in the original equation for extraneous roots.

Example 1. Solve the equation

$\sqrt{x - 1} = 3$ The radical is already isolated.

$(\sqrt{x - 1})^2 = 3^2$

$x - 1 = 9$

$x = 1 + 9$
$x = 10$

Check: $\sqrt{(10-1)} = \sqrt{9} = 3$

Example 2. Solve the equation $\sqrt{x} + 4 = 8$

$$\sqrt{x} + 4 - 4 = 8 - 4$$
$$\sqrt{x} = 4$$
$$(\sqrt{x})^2 = 4^2$$
$$x = 16$$

Example 3. Solve the equation $\sqrt{x} = -9$

$$(\sqrt{x})^2 = (-9)^2$$
$$x = 81$$
Check:
$$\sqrt{81} = -9 \text{ is false.}$$ Thus 81 is not a solution of the equation. The equation $\sqrt{x} = -9$ has no **real solutions**.

Example 4. Solve the equation below.

$$\sqrt{2x - 1} = x - 2$$
$$(\sqrt{2x - 1})^2 = (x - 2)^2$$
$$2x - 1 = x^2 - 4x + 4$$
$$0 = x^2 - 4x - 2x + 4 + 1$$
$$0 = x^2 - 6x + 5$$
$$0 = (x - 5)(x - 1)$$

Thus, there are two possible solutions : if $(x - 5) = 0$, then $x = 5$. If $(x - 1) = 0$, then $x = 1$. A check in the original equation shows that the only solution is $x = 5$.

Example 5. Solve the equation below.

$$\sqrt{5x + 6} - \sqrt{9x - 2} = 0$$
$$\sqrt{5x + 6} = \sqrt{9x - 2}$$
$$(\sqrt{5x + 6})^2 = (\sqrt{9x - 2})^2$$
$$5x + 6 = 9x - 2$$
$$0 = 9x - 5x - 2 - 6$$
$$0 = 4x - 8$$
$$4x = 8$$
$$x = 2$$

A check will show that $x = 2$ is the solution.

Exercise set 9.5

Solve each equation. If the equation has no real solution, so state.

1. $\sqrt{z}=4$ 2. $\sqrt{z}=-9$ 3. $\sqrt{3y}=2$ 4. $4\sqrt{y}=8$

5. $2\sqrt{3y}=6$ 6. $\sqrt{y+3}=5$ 7. $\sqrt{2x-1}=3$ 8. $\sqrt{2x+2}=6$

9. $\sqrt{5y-1}-3=0$ 10. $2=8-\sqrt{y}$

11. $4\sqrt{x+7}-5=11$ 12. $4+\sqrt{x}=2$ 13. $\sqrt{2y-5}=y-4$

14. $\sqrt{x^2+3}=x+1$ 15. $x=4+\sqrt{x^2-32}$ 16. $\sqrt{y+1}=\sqrt{2y-7}$

17. $\sqrt{y^2+3}=y+1$ 18. $\sqrt{3y-2}=\sqrt{y+4}$

19. $\sqrt{2x-4}=\sqrt{x+1}$

Answers to exercise set 9.5

1. z = 16   2. no solution   3. y = 4/3   4. y = 4
5. y = 3   6. y = 22   7. x = 5   8. x = 17
9. y = 2   10. y = 36   11. x = 9   12. no solution
13. y = 7   14. x = 1   15. x = 6   16. y = 8
17. y = 1   18. y = 3   19. x = 5   20. y = 9

9.6 Applications of Radicals

Summary:

> **Pythagorean Theorem**
>
> 1. The square of the hypotenuse of a right triangle is equal to the sum of the squares of the two legs.
>
> 2. If **a** and **b** represent the legs, and **c** represents the hypotenuse, then
>
> $a^2 + b^2 = c^2$.

Example 1. Find the hypotenuse of a right triangle whose legs are 6 feet and 8 feet.

Solution: Use $c^2 = a^2 + b^2$.
$$c^2 = 6^2 + 8^2$$
$$c^2 = 36 + 64$$
$$c^2 = 100$$
$$(c^2)^{1/2} = 100^{1/2}$$
$$c = 10$$

Example 2. The hypotenuse of a right triangle is 17 inches. Find the second leg if one leg is 8 inches.

Solution: Use $c^2 = a^2 + b^2$
$$17^2 = a^2 + 8^2$$
$$289 = a^2 + 64$$
$$289 - 64 = a^2$$
$$225 = a^2$$
$$(225)^{1/2} = (a^2)^{1/2}$$
$$15 = a$$

Example 3. A rectangular athletic field is 50 yards by 100 yards. How long is the diagonal of this field ?

The diagonal forms a right triangle whose legs are 50 yards by 100 yards. Using the Pythagorean theorem and calling the diagonal c, we have $c^2 = a^2 + b^2$. $c^2 = 50^2 + 100^2$. $c^2 = 2500 + 10000$ or $c^2 = 12,500$.

Taking the square root of both sides of the equation, we have $c = \sqrt{12,500}$. $c \approx 111.8$ yards.

Distance Formula

The distance between two points (x_1, y_1) and (x_2, y_2) is given by

$$d = \sqrt{(x_2 - x_1)^2 + (y_2 - y_1)^2}$$

Example 4. Find the length of the line segment which connects the points (2, 3) and (5,7).

Using the distance formula, $d = \sqrt{(x_2 - x_1)^2 + (y_2 - y_1)^2}$

$$d = \sqrt{(5 - 2)^2 + (7 - 3)^2}$$

$$d = \sqrt{9 + 16}$$

$$d = \sqrt{25} = 5$$

Example 5. A rock dropped from a cliff falls a distance of s feet in t seconds according to the formula $s = 16t^2$. If the rock was dropped from a height of 144 feet, how long was the rock in the air?

$$s = 16t^2$$
$$144 = 16t^2$$
$$\frac{144}{16} = t^2$$
$$9 = t^2$$
$$\sqrt{9} = t \quad \text{(we need not worry about the solution } - \sqrt{9} \text{ since time cannot be negative)}$$
$$3 = t$$

Example 6. The formula for the period, in seconds, T, of a pendulum (the time required for the pendulum to make one complete swing both back and forth) is $T = 2\pi \sqrt{\dfrac{L}{32}}$, where L is the length of the pendulum in feet. Find the period of the pendulum if its length is 2 feet. Use 3.14 as an approximation for π.

$T = 2(3.14) \sqrt{\dfrac{2}{32}}$

$T = 6.28 \sqrt{\dfrac{1}{16}}$

$T = 6.28 \cdot \dfrac{1}{4}$

$T \approx 1.57$ seconds.

Exercise set 9.6

Use the Pythagorean theorem to find the quantity indicated.

1. The hypotenuse of a right triangle is 17 inches. One leg of the right triangle is 8 inches. Find the length of the other leg.

2. The two legs of a right triangle are 12 inches and 5 inches. How long is the hypotenuse of this triangle?

3. One leg of a right triangle is 9 inches, while the hypotenuse of this triangle is 15 inches. Find the length of the remaining leg of this right triangle.

4. The two legs of a right triangle are 7 feet and $\sqrt{3}$ feet. Find the length of the hypotenuse. Express answer as a radical.

5. Find the side length of a square whose diagonal is 12 units.

6. The length of a rectangle is 12 cm. Its width is $\sqrt{3}$ cm. How long is the diagonal of this rectangle ? Express answer as a radical.

7. A right triangle is inscribed in a circle so that the hypotenuse of the right triangle is the same as the diameter of the circle.

If the hypotenuse of the triangle is 10 units, and one leg of the triangle is 6 units, find the length of the other leg.

8. Find the diagonal of a square whose side length is 5 units.

9. A 30 foot ladder is leaning against a building. If the base of the ladder is 4 feet from the building, how far up the building does the ladder reach?

10. Find the length of the line segment connected the points (1,5) and (2,-6).

11. Find the length of the line segment connecting the points (4,2) and (-2,4) .

12. Find the side of a square whose area is 230 m².

13. Find the length of the diagonals of a 7 inch x 12 inch rectangle.

14. Find the period of a pendulum if the length of the pendulum is 1 foot. Use $T = 2\pi \sqrt{\dfrac{L}{32}}$

15. The formula for the area of a circle is $A = \pi r^2$, where $\pi \approx 3.14$ and r is the radius of the circle. Find the radius of a circle whose area is 78.5 in².

16. Find the radius of a circle whose area is 153.86 m².

17. How long a wire is needed to reach from the top of a 30 foot pole to a point 5 meters from the base of the pole ?

18. The diagonal of a large rectangular family room is 27 feet. If the room is 10 feet wide, approximately how long is the room?

19. A television has a 19 inch diagonal. If the length is 10 inches, how wide is the television screen, assuming it is rectangular?

20. Find the hypotenuse of a right triangle whose legs are $\sqrt{17}$ and $\sqrt{19}$.

Answers to exercise set 9.6

1. 15 inches   2. 13 inches   3. 12 inches   4. $2\sqrt{13}$ ft.

5. $6\sqrt{2}$ 6. $7\sqrt{3}$ cm   7. 8 units   8. $5\sqrt{2}$

9. 29.73 feet   10. ≈11.05   11. ≈6.32   12. ≈15.17 feet

13. ≈ 13.89 inches   14. ≈ 1.11 seconds   15. r = 5 inches

16. r = 7 meters   17. ≈ 30.4 meters   18. ≈ 25 feet

19. ≈ 16.1 inches   20. 6 inches

9.7 Higher Roots and Rational Exponents

Summary:

$\sqrt[3]{a}$ is read "the **cube root of a**"

$\sqrt[4]{a}$ is read "the **fourth root of a**"

$\sqrt[3]{a} = b$ if $b^3 = a$

$\sqrt[4]{a} = b$ if $b^4 = a$

Example 1. Evaluate the following.

a) $\sqrt[3]{27}$

$\sqrt[3]{27} = 3$ since $3^3 = 3 \cdot 3 \cdot 3 = 27$

b) $\sqrt[3]{125}$

$\sqrt[3]{125} = 5$ since $5^3 = 125$

c) $\sqrt[3]{-27}$

$\sqrt[3]{-27} = -3$ since $(-3)(-3)(-3) = -27$

d) $\sqrt[3]{-125}$

$\sqrt[3]{-125} = -5$ since $(-5)^3 = (-5)(-5)(-5) = -125$.

Example 2. Evaluate the following.

a) $\sqrt[4]{16}$

$\sqrt[4]{16} = 2$ since $2^4 = 2 \cdot 2 \cdot 2 \cdot 2 = 16$

b) $\sqrt[4]{625}$

$\sqrt[4]{625} = 5$ since $5^4 = 5 \cdot 5 \cdot 5 \cdot 5 = 625$

c) $\sqrt[4]{1296}$

$\sqrt[4]{1296} = 6$ since $6^4 = 6 \cdot 6 \cdot 6 \cdot 6 = 1296$

Note: The cube root of a positive number is a positive number and the cube root of a negative number is a negative number. The radicand of a fourth root (or any even root) must be a nonnegative number for the expression to be a real number.

Summary:

Product Rule for Radicals

$\sqrt[n]{a} \cdot \sqrt[n]{b} = \sqrt[n]{ab}$, for $a \geq 0$, $b \geq 0$

Rewrite a radical in exponential form

$\sqrt[n]{a} = a^{\frac{1}{n}}$, $a \geq 0$

$\sqrt[n]{a^m} = a^{\frac{m}{n}}$, $a \geq 0$, and m, n are integers

Example 3.   Simplify the following

a) $\sqrt[3]{40}$

$\sqrt[3]{40} = \sqrt[3]{8 \cdot 5} = \sqrt[3]{8} \cdot \sqrt[3]{5} = 2\sqrt[3]{5}$

b) $\sqrt[3]{54}$

$\sqrt[3]{54} = \sqrt[3]{27 \cdot 2} = \sqrt[3]{27} \cdot \sqrt[3]{2} = 3\sqrt[3]{2}$

c) $\sqrt[4]{48}$

$$\sqrt[4]{48} = \sqrt[4]{16 \cdot 3} = \sqrt[4]{16} \cdot \sqrt[4]{3} = 2 \cdot \sqrt[4]{3}$$

d) $\sqrt[4]{162}$

$$\sqrt[4]{162} = \sqrt[4]{81 \cdot 2} = \sqrt[4]{81} \cdot \sqrt[4]{2} = 3 \cdot \sqrt[4]{2}$$

Example 4. Simplify the following.

a) $\sqrt[3]{y^7}$

$$\sqrt[3]{y^7} = \sqrt[3]{y^6 \cdot y} = \sqrt[3]{y^6} \cdot \sqrt[3]{y} = y^2 \cdot \sqrt[3]{y}$$

b) $\sqrt[3]{a^5 b^9} = \sqrt[3]{a^3 \cdot a^2 \cdot b^9} = \sqrt[3]{a^3 b^9} \cdot \sqrt[3]{a^2} = ab^3 \cdot \sqrt[3]{a^2}$

c) $\sqrt[4]{x^7}$

$$\sqrt[4]{x^7} = \sqrt[4]{x^4 \cdot x^3} = \sqrt[4]{x^4} \cdot \sqrt[4]{x^3} = x \cdot \sqrt[4]{x^3}$$

d) $\sqrt[4]{a^5 b^9}$

$$\sqrt[4]{a^5 b^9} = \sqrt[4]{a^4 b^8} \cdot \sqrt[4]{ab} = ab^2 \cdot \sqrt[4]{ab}$$

Example 5. Evaluate the following.

a) $(25)^{3/2}$

$$(25)^{3/2} = (\sqrt{25})^3 = 5^3 = 125$$

b) $(-27)^{2/3}$

$$((-27)^{2/3} = (\sqrt[3]{-27})^2 = (-3)^2 = 9$$

c) $64^{1/3}$

$$64^{1/3} = \sqrt[3]{64^1} = 4$$

d) $16^{3/4}$

$$16^{3/4} = \left(\sqrt[4]{16}\right)^3 = 2^3 = 8$$

Example 6. Write each of the following in exponential form.

a) $\sqrt[4]{y^{10}}$

$$\sqrt[4]{y^{10}} = y^{10/4} = y^{5/2}$$

b) $\sqrt{x^2 y^5}$

$$\sqrt{x^2 y^5} = (x^2 y^5)^{1/2} = x^{2/2} y^{5/2} = xy^{5/2}$$

c) $\sqrt[3]{a^8}$

$$\sqrt[3]{a^8} = a^{8/3}$$

d) $\sqrt[3]{10a^6} = (10a^6)^{1/3} = \left(10^{\frac{1}{3}} \cdot a^{\frac{6}{3}}\right) = 10^{\frac{1}{3}} \cdot a^2$

Example 7. Simplify the following.

a) $\sqrt[3]{y} \sqrt{y^3}$

$$\sqrt[3]{y} \sqrt{y^3} = y^{1/3} y^{3/2}$$
$$= y^{1/3 + 3/2}$$
$$= y^{2/6 + 9/6}$$
$$= y^{11/6}$$
$$= \sqrt[6]{y^{11}}$$

b) $\left(\sqrt[3]{w^4}\right)^5$

$$\left(\sqrt[3]{w^4}\right)^5 = \left(w^{\frac{4}{3}}\right)^5$$
$$= w^{\frac{20}{3}}$$

$$= \sqrt[3]{w^{20}}$$

$$= w^6 \cdot \sqrt[3]{w^2}$$

Exercise set 9.7

1. $\sqrt[3]{216}$
2) $\sqrt[4]{16}$
3) $\sqrt[3]{-1}$
4. $\sqrt[3]{-8}$

5. $\sqrt[4]{1296}$
6. $\sqrt[4]{-81}$
7. $\sqrt[3]{16}$
8. $\sqrt[3]{250}$

9. $\sqrt[4]{64}$
10. $\sqrt[4]{162}$
11. $\sqrt[3]{135}$
12. $\sqrt[4]{32}$

Simplify

13. $\sqrt[3]{x^5}$
14. $\sqrt[4]{x^5}$
15. $\sqrt[3]{x^{18}}$
16. $\sqrt[4]{a^{25}}$

17. $\sqrt[4]{a^8 b^5}$
18. $\sqrt[3]{x^6 y^{10}}$
19. $(-27)^{\frac{4}{3}}$
20. $(16)^{\frac{5}{4}}$

21. $(-64)^{\frac{2}{3}}$

Write each radical in exponential form.

22. $\sqrt[3]{y^4}$
23. $\sqrt[3]{y^5}$
24. $\sqrt{x^{11}}$

Simplify.

25. $\sqrt{y^3} \cdot \sqrt[3]{y^2}$
26. $\sqrt[4]{y^3} \cdot \sqrt{y}$
27. $\left(\sqrt[3]{x^2}\right)^6$

Answers to section 9.7

1. 6
2. 2
3. -1
4. -2

5. 6
6. not a real number
7. $2\sqrt[3]{2}$
8. $5\sqrt[3]{2}$

9. $2\sqrt[4]{4}$
10. $3\sqrt[4]{2}$
11. $3\sqrt[3]{5}$
12. $2\sqrt[4]{2}$

13. $x\sqrt[3]{x^2}$
14. $x \cdot \sqrt[4]{x}$
15. x^6

16. $a^6 \sqrt[4]{a}$ 17. $a^2 \cdot b\sqrt[4]{b}$ 18. $x^2 y^3 \sqrt[3]{y}$ 19. 81

20. 32   21. 16   22. $y^{\frac{4}{3}}$ 23. $y^{\frac{5}{4}}$

24. $x^{\frac{11}{2}}$ 25. $y^{\frac{13}{6}}$ 26. $y^{\frac{5}{4}}$ 27. x^4

CHAPTER 9 PRACTICE TEST

1. Write $\sqrt{2x}$ in exponential form.

Simplify

2. $\sqrt{(x+4)^2}$

3. $\sqrt{98}$

4. $\sqrt{32x^4y}$

5. $\sqrt{15x^2y}\sqrt{5x^2y^3}$

6. $\sqrt{\dfrac{6}{216}}$

7. $\dfrac{1}{\sqrt{3}}$

8. $\sqrt{\dfrac{9x}{5}}$

9. $\sqrt{\dfrac{60xy^5}{4x^5y^3}}$

10. $\dfrac{3}{1+\sqrt{2}}$

Simplify

11. $6\sqrt{2} - 8\sqrt{2} - \sqrt{2}$ 12. $7\sqrt{40} - 2\sqrt{10}$ 13. $4\sqrt{18} + 5\sqrt{50} + 5\sqrt{32}$

Solve.

14. $\sqrt{3x+1} = 5$ 15. $\sqrt{x} = -3$ 16. $\sqrt{y^2+4} = y+2$

Solve.

17. $\sqrt{3x+5} - \sqrt{5x-9} = 0$

18. A 14 foot ladder leans against a house. If the base of the ladder is 4 feet from the house, how high is the ladder on the house?

19. Find the side of a square whose area is 135 square inches.

20. Find the distance between the points (-2, 5) and (4, 7).

21. Evaluate. $\sqrt[3]{-64}$

Simplify.

22. $\sqrt[4]{162}$ 23. $\sqrt[4]{y^{12}}$

Answers to Chapter 9 Practice Test

1. $(2x)^{1/2}$ 2. $x+4$ 3. $7\sqrt{2}$ 4. $4x^2\sqrt{2y}$

5. $5x^2y^2\sqrt{3}$ 6. $1/6$ 7. $\dfrac{\sqrt{3}}{3}$ 8. $\dfrac{3\sqrt{5x}}{5}$

9. $\dfrac{y\sqrt{15}}{x^2}$ 10. $-3(1-\sqrt{2})$ 11. $-\sqrt{2}$ 12. $12\sqrt{10}$

13. $17\sqrt{2}$ 14. $x=8$ 15. no solution   16. $y=0$

17. $x=7$ 18. ≈ 13.4 feet   19. ≈ 11.6 inches

20. ≈ 6.32 21. -4 22. $3\sqrt[4]{2}$ 23. y^3

10.1 The Square Root Property

Summary

> **Square Root Property**
>
> If $x^2 = a$, then $x = \sqrt{a}$ or $x = -\sqrt{a}$ (abbreviated $x = \pm \sqrt{a}$)

Example 1. Solve the equation $x^2 - 36 = 0$

Solution. In order to apply the square root property, get the variable by itself on one side of the equation.

$$x^2 - 36 = 0$$
$$x^2 = 36$$

Now, apply the square root property

$$x = \pm \sqrt{36}$$
$$x = \pm 6$$

Check:
$$\begin{array}{ll} x = 6 & x = -6 \\ x^2 - 36 = 0 & x^2 - 36 = 0 \\ 6^2 - 36 = 0 & (-6)^2 - 36 = 0 \\ 36 - 36 = 0 & 36 - 36 = 0 \\ \text{True} & \text{True} \end{array}$$

Example 2. Solve the equation $4x^2 = 24$.

Solution. Again, isolate the variable on one side of the equation.

$$4x^2 = 24$$
$$x^2 = 6$$

Now, apply the square root property.

$$x = \pm \sqrt{6}$$

Example 3. Solve the equation $(x + 2)^2 = 5$

Solution.
$$(x + 2)^2 = 5$$
$$x + 2 = \pm \sqrt{5}$$

$$x = \pm\sqrt{5} - 2$$

Thus, solutions are $-2 + \sqrt{5}$ or $-2 - \sqrt{5}$.

Example 4. Solve the equation $(2x - 3)^2 = 16$

Solution.
$$(2x - 3)^2 = 16$$
$$(2x - 3) = \pm\sqrt{16}$$
$$2x - 3 = \pm 4$$
$$2x = \pm 4 + 3$$
$$x = \frac{(\pm 4 + 3)}{2}$$

Thus, solutions are $\frac{-4+3}{2} = \frac{-1}{2}$ or $\frac{+4+3}{2} = \frac{7}{2}$.

Example 5. A rectangular lot is to have an area of 500 square feet. If the length is four times the width, determine the dimensions of the rectangular lot.

Solution.   Let x = width
Let $4x$ = length

Area = length x width
$$500 = 4x \cdot x$$
$$500 = 4x^2$$
$$\frac{500}{4} = x^2$$
$$125 = x^2$$
$$\pm\sqrt{125} = x$$

Since width cannot be negative, width = $\sqrt{125} = 5\sqrt{5}$ feet.

Length = $4(5\sqrt{5}) = 20\sqrt{5}$ feet.

Exercise set 10.1

Solve each equation.

1. $x^2 = 9$
2. $x^2 = 49$

3. $x^2 - 121 = 0$ 4. $9x^2 - 16 = 0$

5. $8x^2 = 32$ 6. $5y^2 = 90$

7. $5a^2 - 50 = 0$ 8. $12y^2 - 48 = 0$

9. $(2x + 1)^2 = 9$ 10. $(3x + 5)^2 = 16$

11. $(x + 3)^2 = 36$ 12. $(t - 7)^2 = 49$

13. $(c + 3)^2 = 11$ 14. $(x - 2)^2 = 18$

15. $(4x + 3)^2 = 81$ 16. $(3x - 7)^2 = 5$

17. The length of a rectangle is three times its width. If the area of the rectangle is 75 square units, find its dimensions.

18. The length of a rectangle is one-half its width. If the area is 72 square units, find its dimensions.

Answers to exercise set 10.1

1. ± 3 2. ± 7 3. ± 11 4. $\pm 4/3$

5. ± 2 6. $\pm 3\sqrt{2}$ 7. $\pm \sqrt{10}$ 8. ± 2

9. $1, -2$ 10. $-1/3, -3$ 11. $3, -9$ 12. $0, 14$

13. $-3 + \sqrt{11}, -3 - \sqrt{11}$ 14. $2 + 3\sqrt{2}, 2 - 3\sqrt{2}$

15. $3/2, -3$ 16. $\dfrac{7+\sqrt{5}}{3}, \dfrac{7-\sqrt{5}}{3}$ 17. 5 by 15

18. 6 by 12

10.2 Solving Quadratic Equations by Completing the Square

Summary

> **To Solve a Quadratic Equation by Completing the Square**
>
> 1. Use the multiplication (or division) property of equality to make the numerical coefficient of the squared term equal to 1.
>
> 2. Rewrite the equation with the constant by itself on the right hand side of the equation.
>
> 3. Take one-half of the numerical coefficient of the first-degree term, square it, and add this quantity to both sides of the equation.
>
> 4. Replace the trinomial with its equivalent squared binomial.
>
> 5. Use the square root property.
>
> 6. Solve for the variable.
>
> 7. Check your answer in the original equation.

Example 1. Solve the equation $x^2 + 2x - 15 = 0$ by completing the square.

Solution. Step 1 does not apply.

Step 2. $x^2 + 2x = 15$

Step 3. $\frac{1}{2}(2) = 1,\ 1^2 = 1$

$x^2 + 2x + 1 = 15 + 1$

Step 4. $(x + 1)^2 = 16$
Step 5. $x + 1 = \pm 4$

Step 6. $x = \pm 4 - 1$

$x = 4 - 1$ or $x = -4 - 1$
$x = 3$ or $x = -5$

Step 7. Check. $x = 3$

$$x^2 + 2x - 15 = 0$$

$$3^2 + 2(3) - 15 = 0$$

$9 + 6 \quad - 15 = 0$
$\quad 15 \quad - 15 = 0$ (True)

Check $x = -5$

$$x^2 + 2x - 15 = 0$$

$$(-5)^2 + 2(-5) - 15 = 0$$

$25 \quad - 10 \quad - 15 = 0$
$\qquad 15 \quad - 15 = 0$ (True)

Example 2. Solve $x^2 - 5x - 24 = 0$ by completing the square

Solution. Step 1 does not apply

Step 2. $x^2 - 5x = 24$

Step 3. $\frac{1}{2}(-5) = \frac{-5}{2}$, $\left(\frac{-5}{2}\right)^2 = \frac{25}{4}$

$$x^2 - 5x + \frac{25}{4} = 24 \; \frac{25}{4}$$

Step 4. $\left(x - \frac{5}{2}\right)^2 = \frac{121}{4}$

Step 5. $x - \frac{5}{2} = \pm \sqrt{\frac{121}{4}}$

$$x - \frac{5}{2} = \pm \frac{11}{2}$$

Example 2 (continued)

Step 6. $x = \frac{5}{2} \pm \frac{11}{2}$

$x = \frac{5}{2} + \frac{11}{2}$ or $x = \frac{5}{2} - \frac{11}{2}$

$x = \frac{16}{2}$ or $x = -\frac{6}{2}$

$x = 8$ or $x = -3$

Step 7. Check $x = 8$

$$x^2 - 5x - 24 = 0$$
$$8^2 - 5(8) - 24 = 0$$
$$64 - 40 - 24 = 0$$
$$24 - 24 = 0 \text{ (True)}$$

Check $x = -3$

$$x^2 - 5x - 24 = 0$$
$$(-3)^2 - 5(-3) - 24 = 0$$
$$9 + 15 - 24 = 0$$
$$24 - 24 = 0 \text{ (True)}$$

Example 3. Solve $3x^2 + 9x = 5$ by completing the square.

Solution. Step 1. $\frac{1}{3}(3x^2 + 9x) = \frac{1}{3}(5)$

$$x^2 + 9x = \frac{5}{3}$$

Step 2 not needed.

Example 3 continued:

Step 3. $\frac{1}{2}(3) = \frac{3}{2}$, $\left(\frac{3}{2}\right)^2 = \frac{9}{4}$

$$x^2 + 3x + \frac{9}{4} = \frac{5}{3} + \frac{9}{4}$$

Step 4. $\left(x + \frac{3}{2}\right)^2 = \frac{47}{12}$

Step 5. $x + \frac{3}{2} = \pm \sqrt{\frac{47}{12}}$

$x + \frac{3}{2} = \pm \sqrt{\frac{141}{6}}$

Step 6. $x = \frac{-3}{2} \pm \sqrt{\frac{141}{6}}$ or $x = \frac{-9 \pm \sqrt{141}}{6}$

Exercise set 10.2

Solve each equation by completing the square.

1. $x^2 - 6x - 16 = 0$
2. $x^2 + 8x - 9 = 0$
3. $y^2 = 8y - 16$
4. $y^2 + 5y = -4$
5. $x^2 + 6x = 5$
6. $x^2 + 4x + 13 = 0$
7. $2x^2 - 7x + 3 = 0$
8. $2y^2 - 3y = 9$
9. $4x^2 + 4x - 3 = 0$
10. $3r^2 - 2r = 2$

Answers to Exercise set 10.2

1. 8, −2   2. −9, 1   3. 4   4. −1, −4

5. $-3 + \sqrt{14}$, $-3 - \sqrt{14}$ 6. no real solution

7. $\frac{1}{2}$, 3   8. $\frac{-3}{2}$, 3   9. $\frac{1}{2}$, $\frac{-3}{2}$

10. $\frac{1+\sqrt{17}}{3}$, $\frac{1-\sqrt{17}}{3}$

10.3 Solving Quadratic Equations by the Quadratic Formula

Summary

> **To Solve a Quadratic Equation by the Quadratic Formula**
>
> 1. Write the equation in standard form $ax^2 + bx + c = 0$, and determine the numerical values for a, b, and c.
>
> 2. Substitute the values a, b, and c from step 1 in the quadratic formula below and then evaluate to obtain the solution.

Quadratic Formula

$$x = \frac{-b \pm \sqrt{b^2 - 4ac}}{2a}$$

Example 1. Solve the equation $x^2 - 3x - 10 = 0$ using the quadratic formula.

Solution. $ax^2 + bx + c = 0$

Step 1 $a = 1,\ b = -3,\ c = -10$

Step 2 $x = \dfrac{-b \pm \sqrt{b^2 - 4ac}}{2a}$

Step 3. $x = \dfrac{-(-3) \pm \sqrt{(-3)^2 - 4(-1)(-10)}}{2(1)}$

$x = \dfrac{3 \pm \sqrt{9 + 40}}{2}$

$x = \dfrac{3 \pm \sqrt{49}}{2}$

Example 1 continued

$$x = \frac{3 \pm 7}{2}$$

$x = \dfrac{3+7}{2}$ or $x = \dfrac{3-7}{2}$

$x = \dfrac{10}{2}$ or $x = \dfrac{-4}{2}$

$x = 5$ or $x = -2$

So, solutions are -2, 5.

Example 2. Solve the equation $x^2 - 6x = -3$ using the quadratic formula.

Solution. $\qquad x^2 - 6x = -3$

Step 1 $\qquad x^2 - 6x + 3 = 0$

$\qquad\qquad ax^2 + bx + c = 0$
$\qquad\qquad a = 1, \ b = -6, \ c = 3$

Step 2 $\qquad x = \dfrac{-b \pm \sqrt{b^2 - 4ac}}{2a}$

$$x = \frac{-(-6) \pm \sqrt{(-6)^2 - 4(1)(3)}}{2(1)}$$

$$x = \frac{6 \pm \sqrt{36-12}}{2}$$

$$x = \frac{6 \pm \sqrt{24}}{2}$$

$$x = \frac{6 \pm \sqrt{4 \cdot 6}}{2}$$

Example 2 continued

$$x = \frac{6 \pm 2\sqrt{6}}{2}$$

$$x = 3 \pm \sqrt{6}$$

So, solutions are $3 + \sqrt{6}$ and $3 - \sqrt{6}$

Example 3. Solve the equation $2x^2 - 3x = -10$ using the quadratic formula.

Solution. $2x^2 - 3x = -10$

Step 1 $2x^2 - 3x + 10 = 0$

$ax^2 + bx + c = 0$
$a = 2, b = -3, c = 10$

Step 2 $$x = \frac{-b \pm \sqrt{b^2 - 4ac}}{2a}$$

$$x = \frac{-(-3) \pm \sqrt{(-3)^2 - 4(2)(10)}}{2(2)}$$

$$x = \frac{3 \pm \sqrt{9 - 80}}{4}$$

$$x = \frac{3 \pm \sqrt{-71}}{4}$$

Since $\sqrt{-71}$ is not a real number, we can go no further. This equation has no real solution. Write **"No real solution"**. Do not write "0" or leave a blank.

Summary

> The expression under the square root sign in the quadratic formula is called the **discriminant**.
>
> $b^2 - 4ac$ **(discriminant)**
>
> **When the discriminant is:**
>
> 1. **Greater than zero**
> $b^2 - ac > 0$, the quadratic equation has **two distinct real number solutions.**
>
> 2. **Equal to zero,** $b^2 - 4ac = 0$ the quadratic equation has **one real number solution.**
>
> 3. **Less than zero,** $b^2 - 4ac < 0$ the quadratic equation has **no real number solution**

Example 4. Find the discriminant of $2x^2 + 3x - 8 = 0$.

Solution. $a = 2, b = 3, c = -8$

$$\begin{aligned} b^2 - 4ac &= (3)^2 - 4(2)(-8) \\ &= 9 - (-64) \\ &= 73 \end{aligned}$$

Since the discriminant is greater than zero, the given equation would have two distinct real number solutions.

Example 5. Without actually finding the solutions, determine which of the following equations have two distinct real number solutions, one real number solution, or no real number solution.

a) $9x^2 - 6x + 1 = 0$

Solution. $a = 9, b = -6, c = 1$

$$\begin{aligned} b^2 - 4ac &= (-6)^2 - 4(9)(1) \\ &= 36 - 36 \\ &= 0 \end{aligned}$$

Since the discriminant equals 0, the equation has one real number solution.

b) $3x^2 + 2x + 12 = 0$

Solution. $a = 3, b = 2, c = 12$

$$b^2 - 4ac = (2)^2 - 4(3)(12)$$
$$= 4 - 144$$
$$= -140$$

Since the discriminant is less than zero, the equation has no real number solutions.

c) $-2x^2 + 5x + 6 = 0$

Solution. $a = -2, b = 5, c = 6$

$$b^2 - 4ac = (5)^2 - 4(-2)(6)$$
$$= 25 - (-48)$$
$$= 25 + 48$$
$$= 73$$

Since the discriminant is greater than zero, the equation has two real number solutions.

Example 6. A rectangle has an area of 30 square feet. If the length of the rectangle is 2 more than the width, determine the dimensions of the rectangle.

Solution. Let x = width

Let $x + 2$ = length

Area = length x width
$30 = (x + 2)(x)$
$30 = x^2 + 2x$
$0 = x^2 + 2x - 30$

Use the quadratic formula with $a = 1, b = 2, c = -30$.

$$x = \frac{-b \pm \sqrt{b^2 - 4ac}}{2a}$$

$$x = \frac{-2 \pm \sqrt{(2)^2 - 4(1)(-30)}}{2(11)}$$

$$x = \frac{-2 \pm \sqrt{4+120}}{2}$$

Example 6 continued.

$$x = \frac{-2 \pm \sqrt{124}}{2}$$

$$x = \frac{-2 \pm \sqrt{4 \cdot 31}}{2}$$

$$x = \frac{-2 \pm 2\sqrt{31}}{2}$$

$$x = \frac{-2}{2} \pm \frac{2\sqrt{31}}{2}$$

$$x = -1 \pm \sqrt{31}$$

The only solution to make sense is $-1+\sqrt{31}$ since width must be positive. Therefore, the dimensions are $-1+\sqrt{31}$ (≈ 4.57 feet) and $-1 + \sqrt{31} + 2 = 1 + \sqrt{31}$ (≈ 6.57 feet)

Exercise set 10.3

Determine whether each equation has two distinct real number solutions, one real number solution, or no real number solution.

1. $x^2 - 3x - 5 = 0$
2. $4x^2 + 12x + 9 = 0$
3. $3x^2 + 2x + 11 = 0$
4. $-2x^2 + 2x + 3 = 0$
5. $x^2 = 6x - 1$
6. $4x = 1 - 3x^2$

Use he quadratic formula to solve each equation. If the equation has no real solutions, so state.

7. $x^2 - 4x - 5 = 0$
8. $t^2 + t - 6 = 0$
9. $2y^2 - y - 1 = 0$
10. $x^2 - 2x + 6 = 0$

11. $t^2 - 2t = 5$ 12. $4x^2 - 8x - 1 = 0$

13. $5x^2 - 6x = 3$ 14. $3x^2 = 5x - 6$

15. A rectangle has an area of 45 square units. If the length of the rectangle is 2 more than twice its width, find the dimensions of the rectangle. Round to the nearest 0.01 inch.

Answers to exercise set 10.3

1. two real number solutions   2. one real number solution
3. no real number solution   4. two real number solutions
5. two real number solutions   6. two real number solutions

7. 5, -1   8. -3, 2   9. $\frac{-1}{2}$, 1

10. no real number solution   11. $1 + \sqrt{6}, 1 - \sqrt{6}$

12. $\frac{2 + \sqrt{5}}{2}, \frac{2 - \sqrt{5}}{2}$ 13. $\frac{3 + 2\sqrt{6}}{5}, \frac{3 - 2\sqrt{6}}{5}$

14. no real number solution   15. width ≈ 4.27 units
                                       length ≈ 10.54 units

10.4 Graphing Quadratic Equations

Summary

> The graph of a quadratic equation in two variables,
> $y = ax^2 + bx + c$, $a \neq 0$
> is called a parabola.
>
> One method that can be used to graph a quadratic equation is to plot it point by point. When determining points to plot, select values for x and determine the corresponding value for y.
>
> When **a** is positive, the parabola will open upward. When **a** is negative, the parabola will open downward.
>
> The **vertex** is the **lowest** point on a parabola that opens upward and the **highest** point on a parabola that opens downward.
>
> The **axis of symmetry** is a line that divides the parabola into two equal halves

Example 1. Graph $y = x^2 + 2$

Since this equation is of the form $y = ax^2 + bx + c$, with a = 1, b = 0 and c = 2, the graph is a parabola. Since a = 1 > 0, the graph will open upward.

	$y = x^2 + 2$	Ordered pair:
Let x = -3	$y = (-3)^2 + 2 = 11$	(-3, 11)
Let x = -2	$y = (-2)^2 + 2 = 6$	(-2, 6)
Let x = -1	$y = (-1)^2 + 2 = 3$	(-1, 3)
Let x = 0	$y = (0)^2 + 2 = 2$	(0, 2)
Let x = 1	$y = (1)^2 + 2 = 3$	(1, 3)

Let x = 2   y = 2² + 2   = 6   (2, 6)

Example 1 continued

Let x = 3   y = 3² + 2   = 11   (3 , 11)

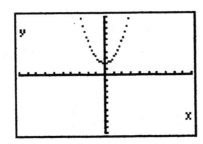

Example 2.   Graph y = -x² + 4x - 4

Solution.   Since a = -1 < 0 , the parabola opens downward

y = -x² + 4x - 4		Ordered pair:
Let x = -1	y = -(-1)² + 4(-1) - 4 = -9	(-1,-9)
Let x = 0	y = 0² + 4(0) - 4 = -4	(0 , -4)
Let x = 1	y = -(1)² + 4(1) - 4 = -1	(1 , -1)
Let x = 2	y = -(2)² + 4(2) - 4 = 0	(2 , 0)
Let x = 3	y = -(3)² + 4(3) - 4 = -1	(3, -1)
Let x = 4	y = -(4)² + 4(4) - 4 = -4	(4, -4)
Let x = 5	y = -(5)² + 4(5) - 4 = -9	(5, -9)

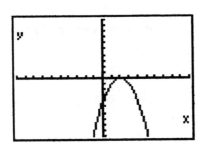

Summary

> **Axis of Symmetry and x coordinate of vertex**
>
> $$x = \frac{-b}{2a}$$
>
> **y coordinate of vertex**
>
> $$y = \frac{4ac - b^2}{4a}$$
>
> One method to use in selecting points to plot when graphing parabolas is to determine the axis of symmetry, a vertical line drawn through the vertex of the graph. Then select nearby values of x on either side of the axis of symmetry. When graphing the equation, make use of the symmetry of the graph.

Example 3. Consider $y = x^2 - 2x - 8$. Once again, this equation is of the form $y = ax^2 + bx + c$, so its graph is a parabola.

a) Find the axis of symmetry.

Solution.

$a = 1$, $b = -2$ and $c = -8$

$$x = \frac{-b}{2a} = \frac{-(-2)}{2(1)} = \frac{2}{2} = 1$$

So, $x = 1$ is the equation of the axis of symmetry.

b) find the vertex of the graph.

Solution. From (a), the x coordinate = 1. For the y-coordinate

$$y = \frac{4ac - b^2}{4a}$$

$$y = \frac{4(1)(-8) - (-2)^2}{4(1)}$$

Example 3 continued

$$y = \frac{-32-4}{4}$$

$$y = \frac{-36}{4} = -9$$

So, the vertex is (1, -9).

c) Graph $y = x^2 - 2x - 8$

Solution. Since a >0, parabola opens upward. We will pick x values nearby the x coordinate of the vertex, which is 1.

```
                                                          Ordered pair
Let x = -1       y = (-1)² -2(-1) - 8 = -5                (-1,-5)
Let x = 0        y = (0)²  -2(0)  - 8 = -8                (0, -8)
Let x = 1        y = -9 (already calculated)              (1, -9)
Let x = 2        y = 2²  -2(2)  - 8    = -8               (2, -8)
Let x = 3        y = 3²  -2(3)  - 8    = -5               (3, -5)
```

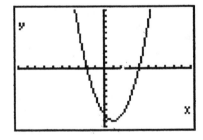

Notice that x values that are the same number of units from the vertex (x = 1) will correspond to the same y value. For instance, when x = -1 (two units to the left of the vertex), y = -5. When x = 3 (two units to the right of the vertex), y = -5. This is what is meant by the parabola being symmetric about the axis of symmetry, x = 1.

Summary

> **X intercepts**
>
> The x intercepts of the graph of the quadratic equation can be found by letting $y = 0$ in the equation $y = ax^2 + bx + c$ and solving for x.
>
> A quadratic equation of the form $y = ax^2 + bx + c$ will have either two distinct x intercepts, one x intercept, (which will be the vertex), or no x intercepts.

Example 4. Find the x intercepts of the graph of $y = x^2 + 3x - 10$ by factoring, completing the square and the quadratic formula.

Solution.   Method 1 (Factoring)

$$x^2 + 3x - 10 = 0$$
$$(x + 5)(x - 2) = 0$$
$$(x + 5) = 0 \text{ or } (x - 2) = 0$$
$$x = -5 \text{ or } x = 2$$

Method 2 (Completing the square)

$$x^2 + 3x - 10 = 0$$
$$x^2 + 3x = 10$$

$$\left(\tfrac{1}{2}(3) = \tfrac{3}{2}, \left(\tfrac{3}{2}\right)^2 = \tfrac{9}{4} \right)$$

$$x^2 + 3x + \tfrac{9}{4} = 10 + \tfrac{9}{4}$$

$$\left(x + \tfrac{3}{2}\right)^2 = \tfrac{49}{4}$$

$$\left(x + \tfrac{3}{2}\right) = \pm \tfrac{7}{2}$$

$$x = \tfrac{-3}{2} \pm \tfrac{7}{2}$$

Example 4 continued

$x = \frac{-3}{2} + \frac{7}{2}$ or $x = \frac{-3}{2} - \frac{7}{2}$

$x = \frac{4}{2}$ $\quad\quad\quad\quad\quad\quad\quad\quad x = \frac{-10}{2}$

$x = 2$ $\quad\quad\quad\quad\quad\quad\quad\quad\quad x = -5$

Method 3 (Quadratic formula)

$$x^2 + 3x - 10 = 0$$

$$a = 1, \ b = 3, \ c = -10$$

$$x = \frac{-b \pm \sqrt{b^2 - 4ac}}{2a}$$

$$x = \frac{-3 \pm \sqrt{(3)^2 - 4(1)(-10)}}{2(1)}$$

$$x = \frac{-3 \pm \sqrt{9 + 40}}{2}$$

$$x = \frac{-3 + 7}{2}$$

$x = \frac{-3 - 7}{2}$ or $x = \frac{-3 + 7}{2}$

$x = \frac{-10}{2}$ $\quad\quad\quad\quad\quad\quad\quad x = \frac{4}{2}$

$x = -5$ $\quad\quad\quad\quad\quad\quad\quad\quad x = 2$

Example 4 (b) Graph $y = x^2 + 3x - 10$

Solution. Find x coordinate of the vertex.

$$x = \frac{-b}{2a} = \frac{-3}{2(1)} = \frac{-3}{2}$$

Find y coordinate of vertex.

$$y = \left(\frac{-3}{2}\right)^2 + 3\left(\frac{-3}{2}\right) - 10 = -12\frac{1}{4}$$

Vertex is $\left(\frac{-3}{2}, -12\frac{1}{4}\right)$

Since $a = 1 > 0$, parabola opens upward.

```
                                                    Ordered pair
Let x = -4 , y = (-4)² + 3(-4) - 10 = -6            (-4,-6)
Let x = -3 , y = (-3)² + 3(-3) - 10 = -10           (-3, -10)
Let x = -2 , y = (-2)² + 3(-2) - 10 = -12           (-2, -12)
Let x = -1 , y = (-1)² + 3(-1) - 10 = -12           (-1, -12)
Let x =  0 , y = (0)²  + 3(0)  - 10 = -10           ( 0, -10)
```

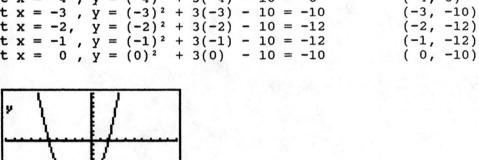

Exercise set 10.4

Indicate the axis of symmetry, the coordinates of the vertex, and whether the parabola opens up or down.

1. $y = x^2 + 4x - 3$
2. $y = x^2 - 10x + 12$
3. $y = 2x^2 + 4x$
4. $y = 3x^2 - 15$
5. $y = -3x^2 + 12x - 6$
6. $y = 4 - x - 2x^2$

Graph each quadratic equation and determine the x-intercepts if they exist.

7. $y = x^2 + 2$
8. $y = x^2 - 4x + 10$
9. $y = x^2 + 2x - 15$
10. $y = x^2 - 6x + 8$
11. $y = x^2 - 4x + 9$
12. $y = 3x^2 + 6x + 3$

Answers to exercise set 10.4

1. $x = -2$, $(-2, -7)$, upward
2. $x = 5$, $(5, -13)$, opens upward
3. $x = -1$, $(-1, -2)$, upward
4. $x = 0$, $(0, -15)$, opens upward
5. $x = 2$, $(2, 6)$, opens downward
6. $x = \frac{-1}{4}$, $\left(\frac{-1}{4}, \frac{33}{8}\right)$

7.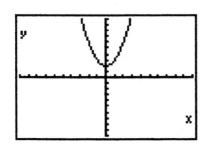

8.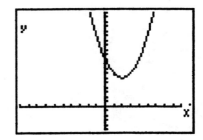

9.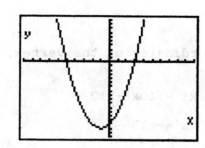
10.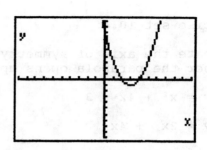
11.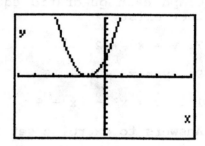
12.

Chapter 10 Practice Test

1. Solve the equation $x^2 - 4 = 22$

2. Solve the equation $(3x + 2)^2 = 15$

3. Solve by completing the square: $x^2 + 4x = 21$

4. Solve by completing the square: $x^2 - 3x - 18 = 0$

5. Solve by the quadratic formula: $2x^2 - 5x - 3 = 0$

6. Solve by the quadratic formula: $3x^2 - x = 1$

7. Solve by the method of your choice: $2x^2 - 3x - 2 = 0$

8. Solve by the method of your choice: $2x^2 - 4x - 5 = 0$.

9. Determine if the following equation has two distinct real solutions, one real solution, or no real solutions.

$$3x^2 - 6x + 3 = 0$$

10. Indicate the axis of symmetry, the coordinates of the vertex, and whether the graph opens upward or downward: $y = -x^2 - 5x + 3$.

11. Graph the following equations, and determine the x intercepts if they exist: $y = x^2 - 4x + 3$.

12. Graph the following equation, and determine the x intercepts if they exist: $y = x^2 + 4x + 7$.

13. The length of a rectangle is 2 more than three times its width. Find the length and width of the rectangle if its area is 20 square feet.

Answers for Chapter 10 Practice Test

1. $x = \pm\sqrt{26}$

2. $\dfrac{-2 + \sqrt{15}}{3}, \dfrac{-2 - \sqrt{15}}{3}$

3. $-7, 3$

4. $-3, 6$

5. $-1/2, 3$

6. $\dfrac{1 + \sqrt{13}}{6}, \dfrac{1 - \sqrt{13}}{6}$

7. $-1/2, 2$

8. $\dfrac{2 + \sqrt{14}}{2}$, $\dfrac{2 - \sqrt{14}}{2}$ 9. one real solution

10. $x = \dfrac{5}{2}$, $\left(\dfrac{5}{2}, \dfrac{-63}{4}\right)$, opens down

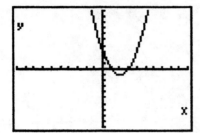
11.

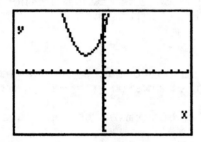
12.

13. width ≈ 2.27 ft.
    length ≈ 8.81 ft.